A New Kind of Mind

A Potential Path toward

Artificial General Intelligence

Glenn Graham

Staghorn Ridge Publications

Printed in the United States by Staghorn Ridge Publications, Harrison, ID 83833.

For information, please email Staghorn Ridge Publications at
staghornridgebooks@gmail.com

FIRST EDITION

Library of Congress Cataloging-in-Publication Data
Name: Glenn Graham, author
Title: A New Kind of Mind: A Potential Path toward Artificial General Intelligence/
Glenn Graham
Description: First Edition: Staghorn Ridge Publications, 2026
Includes bibliographical references
Identifiers: LCCN 2026915917
ISBN 979-8-9966800-0-9 (hardcover)
ISBN 979-8-9966800-1-6 (ebook)
Subjects: LCSH: Artificial Intelligence

Jacket images: first five by Dave Carlson/www.carlsonstockart.com
Last image by Glenn Graham

Artificial Intelligence was not used in the production of this book.

For my wife, life partner, and the love of my life,

Carol

Any Entity that can

Perceive the Present,
Remember and Learn from the Past,
and Contemplate and Predict the Future

is a Conscious and Intelligent Being.

And if that Entity knows that itself is the one
doing the Perceiving, Remembering, Learning,
Contemplating, and Predicting,

it is also Self-Aware.

Contents

A New Kind of Mind

1

Introduction

"The development of full artificial intelligence could spell the end of the human race....It would take off on its own, and re-design itself at an ever-increasing rate. Humans, who are limited by slow biological evolution, couldn't compete, and would be superseded."

- Stephen Hawking

"I visualize a time when we will be to robots what dogs are to humans, and I'm rooting for the machines."

- Claude Shannon

"Some people call this artificial intelligence, but the reality is this technology will enhance us. So instead of artificial intelligence, I think we'll augment our intelligence."

- Ginni Rometty

"Artificial intelligence will reach human levels by around 2029. Follow that out further to, say, 2045, we will have multiplied the intelligence, the human biological intelligence of our civilization a billion-fold."

- Ray Kurzweil

"By far, the greatest danger of artificial intelligence is that people conclude too early that they understand it."

- Eliezer Yudkowsky

Significant progress has been made in the development of Artificial Intelligence (AI) to carry out a wide variety of tasks. Until recently, most of these systems have been designed to accomplish specific tasks, such as voice and visual recognition, natural language processing, disease diagnosis, data processing, games, etc. These forms of AI are sometimes referred to as Artificial Narrow Intelligence (ANI) and can provide significant capabilities for an extensive number of applications.

Many of these ANI systems have had significant commercial success and have inherent value in their own right. However, probably the ultimate goal for most people working in AI is the development of Artificial General Intelligence (AGI). AGI, which involves a range of capabilities that include self-learning, reasoning, predicting, planning, and communicating, in a manner that mimics human abilities, has been an elusive goal.

However, very recently, with the advent of OpenAI's ChatGPT, followed by the other LLMs (large language models), such as Anthropic's Claude, Google's Gemini, Meta's LLaMA, and xAI Grok, major steps are being made in that direction. These models are able to respond to text prompts with impressive results that demonstrate amazing capabilities, such as summarize information from the internet, a report, or a meeting; create documents including job application letters and resumes; create complex computer code; analyze data; provide answers to microbiology and MBA exams; write poems about a modern subject written in the same style as William Shakespeare; and create procedures for laboratory experiments, just to mention a few.

But, LLMs do not "think" in the same way as a human. They do not understand the meaning of the words or concepts that they are presenting and they certainly are not aware or conscious of themselves or what they are doing.

In addition, these models often hallucinate and make very convincing statements that are incorrect. The style, organization, and content of their responses, although impressive, are often not at the level of a true expert in any given field.

LLMs are improving rapidly and are entering a new phase in which they operate as agents, allowing them to control various computer applications. Many people, especially the upper managers of the LLM software companies, believe that these models are on the verge of creating massive unemployment for almost any job that is mainly computer-based.

There is also a lot of discussion, by many of these same people, about how close LLMs are to achieving AGI. The beliefs range from thinking that LLMs have a long way to go, to thinking that they are almost here, depending on how AGI is defined. But many of the AI researchers do think that LLMs will eventually achieve AGI.

However, a number of notable researchers believe that LLMs are not the correct type of model to reach AGI and that a different approach will be required, or that LLMs may eventually achieve AGI, but that the current state of LLMs is a long way from being able to demonstrate AGI capabilities.

Andrew Ng believes that we are decades away from AI achieving AGI based on the definition of AI that "can perform any intellectual task that a human can" [1]. According to Gary Marcus, deep learning, on which LLMs are based, lack the ability to represent causal relationships, cannot perform logical inferences, and are unable to integrate abstract knowledge [2]. Yann LeCun, who was instrumental in the development of LLMs and served as chief AI scientist at Meta for more than a decade, before leaving Meta recently, now believes that LLMs are limited in how powerful they can become, and that a new approach is needed, such as one based a world model [3] [4].

I believe that for an AI to achieve AGI, it needs to think essentially like a human. The generally accepted definition of AGI is that it "matches or surpasses human capabilities across virtually all cognitive tasks [5]." Of course, there would be some differences that would be inherent and others that would depend on the application of the AI. For example, an AI would not have a desire to eat or drink but may instead need to find a charging station or outlet once in a while. And, depending on the type of employment of the AI, it may not need the sense of smell or taste, but may need sensory inputs for ultraviolet light or radio waves. But the main types of cognitive abilities should be very similar.

Therefore, I think that it makes sense to discuss what are the main types of cognitive functions of a human mind so that we know what we would like an AI to imitate. One could disagree as to whether this is the complete list of human cognitive functions and about some of the suggested concepts, but I believe that it does cover most of the main points.

The physical world consists of Entities and their Actions. These 3D Entities are people, animals, plants, rocks, mountains, streams, rivers, lakes, oceans, cars, bicycles, buildings, etc. that are moving, rotating, being constructed, aging, being disassembled or destroyed, growing, shrinking, appearing, disappearing, etc. To understand the physical world, one needs to understand Entities and what they do using the five senses, and the human mind has evolved to do this exceptionally well.

As a first step toward this understanding, the human mind (with the help of its eyes) sees and divides its surroundings into a menagerie of diverse 2D Entities, which are then converted into mental 3D representations of Entities based on comparison with the memories of similar Entities, and the perceived 2D views of a given Entity as it moves and rotates relative to the other Entities in the scene. Tactile sensory input of Entities can also add to visual perception to aid in the formation of the 3D representations.

Entities tend to have Actions and motions that are specific to that Entity. Horses run, bean seeds sprout, snow falls, rivers flow, humans walk, cars drive, and volcanos erupt in ways specific to each of these Entities. The human mind comprehends motion in general, but also the association just described. If asked about a type of Entity that we are familiar with, we would likely be able to describe the types of motions that that Entity is able to perform.

Entities are defined by their Attributes, such as size, shape, weight, volume, temperature, color, sound, texture, smell, taste, etc. And most Entities are composed of assembled parts, or SubEntities. We humans not only rely on these Attributes to determine the type of Entity, but as a way to distinguish between Entities of the same type. For example, John has black hair, weighs 165 lb, and is 5'-10", whereas Mark has blond hair, weighs 190 lb, and is 6'-3". We also are able to understand that engines, washing machines, computers, animals, airplanes, bridges, buildings, etc. are composed of assembled parts, and many of us can comprehend how the subparts, or SubEntities, are assembled, how to assemble and disassemble them, the reason for each of the SubEntities, and how they move relative to each other. And we understand that the SubEntities also have their own Attributes.

The association and interaction between Entities is as important as the Entities themselves. Groups of Entities might involve organization structures (such as the employees in a business), taxonomic hierarchies, animal herds, crowds, automobile traffic, raindrops, stacks of documents, blades of grass, cars in a parking lot, etc. As evident, these associations might be conceptual or physical, have physical interactions, and movement of the Entities relative to each other. The human mind is not only able to understand these relationships, but can create Groups of existing or novel types, and can predict the result of many of their interactions. For example, we know two Entities cannot occupy the same space at the same time, the effect of gravity's attraction between Earth and the Entities present on its surface, movement of Entities relative to each other, and the effect of impact between two Entities, to list a few examples.

Humans understand location, which is usually thought of as being 2D, when we think in terms of a map or location in a city. But it becomes 3D for multi-story buildings, airplanes, and spacecraft. It can also be absolute or relative, as for an address or a location 3 miles away.

Time is a concept that we humans comprehend in numerous ways. Of course, we think about absolute and relative time, as in 2:00 pm today or 5 hours from now. But we also work with time as chronological order, the beginning and ending of Events, motion, rates, seasons, cycles, aging, and reactions, to mention a few. We also have a general overriding concept of time as past, present, and future.

Humans think of the combination of Entities and Actions as Events, such as a car driving to town, a person walking to work, a horse running in a field, an explosion, a sunset, a chemical reaction in a laboratory, a marriage, a party, a symposium, etc. Events involve location and time, and essentially represent who, what, when, and where for one or more Entities and an Action taking place at one or more locations at a specific time or over some time period. Events can be mentally ordered (as in a recipe), organized, rearranged, or revised as if they were a type of object. Groups, as with Entities, can also be part of Events.

The concept of Self is inherent to humans, but it is a complex concept for a person to fully understand. A person tends to think of its own Self as the Entity that is doing his/her perceiving, remembering, thinking, planning, and acting. But the Self is also perceived to interact with other Entities as any other human might do. The Self can act on other Entities, as these other Entities can act on it. So, the Self must understand that it is responsible for its own Actions and reflect on them for self-evaluation, it must protect and work to improve itself, and it must have a chronological memory of its own thoughts and activities so that it can reflect back on these.

One of the key traits of humans is their ability to reason and use logic. Analogy may be one of the most important forms of logic because it allows understanding of an unknown by making a comparison with something that is known. It applies to Entities, Attributes, Groups, Actions, and Events, and is essentially a type of generalization in that it emphasizes the similarities of the key aspects while minimizing the other features. For example, we perceive different models of cars as cars because they all have four tires, a passenger compartment, windshield, and are in a certain size range, even though they have somewhat different shapes, colors, styles, and number of doors. Because of analogy, we can recognize someone we know, even though they are wearing different clothes and standing with a pose we have not seen before, we are able to drive a rental car model we have never driven before, and because we have baked other cakes, we are able to follow a recipe to bake a type of cake we have never made before.

The simplest form of causation, or cause-and-effect, is the occurrence of one Event (cause) which induces a second Event (effect) to happen. Causal relationships can be very complex, involving multiple steps, or Events (some possibly in parallel to each other) such as for a cooking recipe. They can also involve Entities. Causal relationships can often be learned quickly, such as when a glass is dropped on the ceramic tile of the kitchen floor. As we grow up, we learn thousands of cause-and-effect connections, some simple such as when the light switch is flipped, the light turns on, and when you fall, it's going to hurt; others more complicated such as following a procedure or recipe. On a daily basis, we make use of hundreds of causal relations: when the shower faucet is turned on, the water will flow; if I use a towel, it will remove the water from my body, if I put bread in the toaster and turn it on, the bread will become toasted; if I turn the ignition key, the engine will start, etc. However, causation relies on analogy because every time that one of these relationships is repeated, the cause-and-effect Events will each be a little different. We will grab the faucet handle somewhat different each time, we will hold the towel slightly differently and we will dry our body to a different degree, the bread will toast a bit differently each time because the bread slices are not exactly the same, etc. Causation is actually a type of pattern because we see, repeatedly, every time it rains, that the grass gets wet, for example. And causation allows us to predict the future. We can predict that if we turn the faucet handle, that water will come out of the spout, if it rains, the grass will get wet, if we turn the ignition key, the car will start, and if we follow the recipe correctly, we will make a cake that will be delicious (we hope).

A pattern, in terms of logic, is a repeating arrangement or relationship of numbers, shapes, or Events that allows the prediction of the next element either as a repetition of the arrangement or due to a predictable change. This allows for interpolation to be used to determine missing elements or extrapolation to be used to predict pattern elements outside the range of available data. Examples of patterns would be the set of all integers (1, 2, 3, 4…), a repeating set of letters (ABCABCABC…) the Fibonacci sequence in which each element is equal to the sum of the two preceding elements, a repeating geometric pattern, music, the seasons, the day/night cycle (over a short period of days), ocean tides, the life cycle of organisms, and cause-and-effect relationships. Often, analogy is needed to understand and use patterns if the pattern elements have some variation, such as with the last two examples, or for the use of patterns to interpolate and extrapolate, which might include the prediction of future occurrences of Events.

The many disciplines of mathematics are forms of logic that use calculation to allow new information to be determined from existing information. Some of the main branches are algebra, geometry, trigonometry, calculus, statistics, and probability, which allow for the understanding of Entities and Events, and how they will change, and are often needed to help recognize patterns. As with causation and patterns, mathematics allows humans to make predictions.

We use imagination in numerous ways; to visualize the assembly of an Entity such as a model, an engine, a bicycle repair, change of room furniture or décor, what an upcoming vacation will be like, the creation of a document, and thinking ahead for the next steps of a procedure or recipe or to rearrange the steps. When we read a novel, we use our imagination to visualize the scenes described in the book. Imagination is needed for analogy to relate known Entities, Groups, Actions, and Events to analogous ones, to relate a known causal relationship to a new situation, and to interpolate and extrapolate patterns.

Humans have needs that extend from the very basic, such as water, food, and warmth, to self-esteem and feeling important, to aesthetic and self-actualization needs. Emotions represent how we feel about whether these needs are being met. If we are getting enough to eat, we're warm, everything is beautiful around us and we perceive that we have accomplished a lot, we will likely feel happy, elated, and satisfied, and we will have the desire to keep doing whatever we did to reach this state of being. But, if the opposite is true, we will be sad, depressed, and angry, and we will have the desire to make changes that will turn our lives around. These desires form the basis for our goals, plans, and actions.

Language serves two main purposes for humans. The more obvious purpose is to allow us to communicate with each other. However, it also promotes thinking by encapsulating concepts into a single word. Instead of saying "that large four-legged animal, with hooves, a long full tail, a mane on its neck, short hair, a range of colors, that people ride for transportation," we can just say "horse." So, we can think about complex Entities or Events using a minimum of words, each one encapsulating all of the details inherent in a given Entity or Event. Think about the following sentence, "The horse was running across the expansive meadow, with a backdrop of snow-capped mountains, on a warm, mostly sunny, day." Likely, a short video played in your mind of the Event, in which your imagination filled in all of the many details based on your own knowledge and experiences. With just a few words, a relatively complex scene has been created in your mind. And if we speak this same sentence to another person, we are transferring a tremendous amount of information with just a few words (the images retrieved from memory by the other person will, of course, be somewhat different based on their own experiences, but should be conceptually similar).

Long-term memories are stored in the human brain based on various organizational structures that include hierarchies, semantic networks, and schemas, to mention a few. However, let's consider how we personally remember our experiences. We are able to remember the many thousands of Entities and their Attributes that we have encountered throughout our lives, including various versions based on different views, changes due to the seasons or, for humans, changes of clothing and hair styles, and transformations caused by time (semantic memory). We also are aware of the types of Actions associated with many or most of these Entities. And we maintain massive mental files of many of the Events for these Entities, which comprise not only the Entity, but the Action, time, and location of the Events (episodic memory). And these Events are usually strung together in a series (sometimes in parallel) that represents what we experienced during the last hour, day, year, or a lifetime, and we also have a personalized version of this for the experiences of the Self, which include the needs, emotions, and desires that we feel and think about for each of the Self-based Events. We know about thousands of Groups of numerous types and the relationship between the Group members (Entities), whether hierarchical, and physical or conceptual. And we have memories involving language, including words, grammar, and its relationship to Entities, Actions, and Events. We are also able to remember the logical relationships discussed above.

The human brain is able to make goals based on needs, emotions, and desires, to make predictions using the logical relationships presented earlier, to plan and execute actions, then to evaluate the results to determine if the goal was satisfied. In this way, we have the incentives and capabilities to avoid situations that are unpleasant or harmful and migrate toward situations that are desirable and necessary for human existence.

We experience the present, including the associated emotions. We remember the Entities and what has occurred in the past, we evaluate our Actions, and we use this knowledge to make or improve logical relationships. We contemplate and predict the future using these logical relationships to carry out Actions that maintain and improve our lives.

Throughout this book, I expand on these, and additional, concepts to elucidate the kinds of thoughts and mental activities that occur inside of our human brains. The goal here is not to explain how the brain works, but rather to understand what the types of thoughts that we have are, and what we think about on a daily basis.

But the ultimate goal of this book is to provide insights about the creation of an AI with Artificial General Intelligence (AGI). In order to do this, we need to understand what are the types of things that we, as humans, think about, and then implement mental functions in an AI that will produce the same kinds of thoughts. Of course, there are some mental activities for us that would not pertain to an AI, for example, those that are related to human biology. An AI will not have any needs relating to eating or drinking, but, instead, would have a need for electrical power or plugging into a charger periodically. Also, an AI may not need to have all five senses that a human has depending on the type of job the AI would be doing (maybe smell and taste are not needed), but it may need additional senses, such as for ultraviolet, infrared, or ultrasound. And an AI would not have our same emotions but may have some type of artificial emotions that serve a similar purpose to that for humans of encouraging the avoidance of bad things and the pursuit of good things.

Although the AI would need to be able to carry out most of the same mental activities as a human brain, it does not necessarily need to be able to do it in exactly the same way.

So, how similar would an artificial brain need to be to a human brain to match the majority of its capabilities? I believe that the approach taken by the Wright brothers to develop the airplane can provide some insight.

During the time that Orville and Wilbur Wright were developing powered flight, they studied birds to gain insights into the main forces involved. These forces are thrust, lift, drag, and weight, in addition to the forces affecting rotation about the three axes. The wings were the key area of concern. To create sufficient lift, the Wright brothers knew that the wings had to be curved in the same manner as a bird (they performed experiments to define the exact dimensions). However, they knew that they could not create thrust in the same manner as a bird, by flapping its wings, so they developed a propeller (using boat propellers as a model) and designed an engine with a high power-to-weight ratio.

To deal with rotation about the axis of the fuselage, the Wright brothers observed that birds twist, or warp, their wings when they want to rotate about this axis. So, the Wright brothers developed a method of warping the plane's wings to copy what birds do. This worked well initially but later became impractical as the size of planes (and their wings) increased. Ailerons (small moveable surfaces on the trailing outer edge of the wing) replaced wing-warping rather quickly.

Powered flight in heavier-than-air craft was the goal of the Wright brothers (and several other inventors working separately on the problem). They borrowed ideas from birds, but they did not exactly copy everything about a bird, and then just scale it up. They only copied what made sense. Even the wing-warping concept was eventually rejected in favor of a solution that was more applicable to a large man-made machine. The key point here is what concepts from birds needed to be copied exactly and what concepts could, or should, be implemented in a different, more effective, manner.

I believe that if the concepts presented in this book can be implemented in an AI, that it will not only achieve AGI, but that it will be conscious and self-aware. But to the same degree as a human? I don't think that it's possible to predict this yet. However, I think that there is the possibility that an AI could be even more conscious than a human if the sensory input has higher resolution, a larger fraction of the input can be stored in memory, and more connections between stored information can be made by the AI.

I will also explain why I believe that an AI that achieves AGI based on the concepts in this book will immediately achieve ASI (Artificial Super Intelligence).

And I will explain why I don't believe that an AI that achieves AGI, or ASI, would be inherently dangerous to the human race. However, it could be a danger if evil incentives are purposely programmed into the AI software. This point highlights the need to advocate for regulations that will prevent the creation of AIs that have bad intentions for humans and the planet.

The ideas presented in this book try to represent what the human brain is doing, or appears to be doing, at various levels, on a conceptional basis, without trying to explain how the human brain is actually doing it. Where possible, references are made to previous work in each area that is discussed. However, many of these ideas are based on observations of what we, as humans, think about, and the reverse engineering of our thought processes, especially those involved with reasoning and prediction.

It is my belief that, if the concepts presented here can be implemented in hardware and software, an AI with AGI (and ASI), and consciousness and self-awareness, is feasible.

2

Entities, Actions, and Events

"Most people spend most of their time in life moving things from one place to another."

- Glenn Graham

The cognitive part of the human brain spends most of its time trying to understand Entities and what they do, predict what they will do, and remember what they have done.

If you think about your normal activities, they involve getting yourself up in the morning, going to the kitchen, assembling a breakfast, putting clothes on, going to work, going to the store to get groceries off the shelves and put them your cart, then in your car to take home, watching your children, dog, other animals, insects, flowers, trees, clouds, remembering what you did yesterday, planning what you will do tomorrow…

The list is endless, but all of these activities involve passively observing Entities doing something or actively making Entities do something, remembering what happened or what we did previously, and predicting what is going to happen.

Now, let's go into more detail, starting with Entities, then Actions, and then how these two concepts are combined to create Events.

Entities

Every known thing is some kind of Entity, so this term refers to animals, plants, minerals…everything. Entities, if referred to grammatically, are essentially nouns. The following sections discuss some of the main characteristics of Entities.

Dimensions

All Entities, in reality, have 3 dimensions, but some are perceived as having only 2, 1, or even 0 dimensions.

So, people, animals, rocks, trees, cars, would normally be seen to have 3 dimensions with a certain volume. However, Entities such as letters on a page, photos, drawings, maps, patterns on a surface, a chess board, are perceived as only having 2 dimensions, which cover a certain area. A paper photo actually does have some thickness, but it is very thin, and the focus is on the contents of the photo, so that the brain only registers the 2-dimensional space.

A line on a page, a thin stick, a wire coat hanger, or a river or border on a map are perceived as line with only 1 dimension. A star in the sky, period on a page, or a small grain of sand are seen as a point with 0 dimension.

Of course, when we look more closely at these Entities that we perceive as only having 2, 1, or 0 dimensions, we see that they all do actually have 3 dimensions or are images of Entities that have 3 dimensions.

And, in general, the human brain expects that most Entities have 3 dimensions. In fact, the brain appears to have a wired-in system to mentally create 3D images and relies on the senses, mainly sight and touch to fill in the details of these images and their 3D aspects. Most people take this for granted and mainly use sight to identify and define the cars, trees, dogs, signposts, buildings, other people, and other Entities around them. However, people without sight are still able to create 3D images in their brains to represent real Entities around them by using their sense of touch [6]. This allows them to navigate inside and outside of their homes, organize and store their belongings, acquire things in stores, etc. So, even without ever having the sense of sight, blind people are able to understand 3D space in about the same way as a sighted person. And sighted people are also able to use touch to identify Entities (for a simple example, when you go into a dark room and feel the wall for the light switch, you can identify it and its shape so that you know how to move it to switch it on). Therefore, the concept of 3D Entities is one that is very fundamental to the human brain.

State of Being

State of Being represents the physical state or type of Entity. Two examples are:

- Solid-Liquid-Gas
- Inanimate-Plant-Animal

One of the main characteristics of a solid is that it maintains its shape. Liquid flows according to gravity (normally) and is often contained in a vessel or by shores or a riverbank. Gases are not very much influenced by gravity and will spread in all directions. They can be compressed and stored in containers. Also, gases are often invisible (air). The physical state will influence the type of Actions that an Entity can have and the way in which it interacts with other Entities.

Inanimate Entities do not move or change unless influenced by some external force or energy. Plants do not move on their own, but they grow and will change according to the seasons. Animals move on their own, interact with other animals, breathe, eat, drink, etc.

Living-Dead are a sub-category for Inanimate-Plant-Animal. A dead plant or animal is a type of Inanimate Entity, except that it was, at one time, living and was categorized as a plant or animal. So, the Living-Dead designation for plants and animals provides a long-term understanding of them and these types of Entities.

Existence

The human brain understands the concept of Existence. This is somewhat abstract and deals with the past, present, and future. There are 4 main Existence categories:

- Exists
- Existed
- Imaginary
- Non-Existent

Possibly, there are more, but I believe that these 4 cover the main categories. "Exists" simply indicates that the Entity currently exists. It may or may not have existed in the past, but the main point is that the Entity exists in the present time. A person may know the time period for which another person, a dog, a car, etc., has been in existence, or they may have just seen it, know that it currently exists, but not have any information on its history. The same would apply for an AI: it may or may not know the existence history. However, it should have the capability to store this information if it is available.

"Existed" means that the Entity existed in the past but does not currently exist. If the beginning and ending times of existence are known, and the ending time is prior to the current time, then it is obvious that this Entity would be in the category of "Existed."

The category "Imaginary" means that the Entity never has existed and has only been created in the imagination. It could be that the person or AI is designing a new product or determining what it would be like if an Entity of this type was in this location. Or it could be that an Entity is expected to appear or come into existence and would be imagined in advance.

The "Nonexistent" category is a bit complex, in that it deals with Entities that could exist, but do not. For example, for the statement, "He did not have a scar on his hand," there is no scar, with the scar being the Entity of interest here. To fully understand this sentence, a person or AI needs to be able to visualize the image of a scar but take note that one does not actually exist in this case. For an AI that reads or hears this sentence and has knowledge of the "He" referred to here, this knowledge may not include any information about a scar. However, this lack of knowledge about a scar does not necessary mean that there is no scar. It is possible that "He" has a scar, but the AI is just not aware of it. So, it is important that the database for a given Entity ("He" in this case) can record that certain aspects definitely are not part of this Entity if information to this point is known.

A set of subcategories is "Living-Dead." "Living" would correspond with the category "Exists." Whereas "Dead" would tend to relate to the category "Existed." However, if the Entity had just died, its body could still exist, but the living part of the Entity might be referred to as "Existed." At some point in time, the dead Entity would be categorized as "Existed."

A set of super-categories for Existence is "Real-Fictional." The current discussion has generally pertained to "Real" Entities. However, these categories would also apply to Fictional works, such as novels and fairy tales. A Fictional story could be set in present, with most of the characters categorized as "Exists." However, there could be a character that, in the story, had previously died, so would be in the category "Existed." Similar discussions could be made about the "Imaginary" and "Nonexistent" categories.

Real-Fictional

An Entity can be Real or Fictional. A Real Entity is one that actually Existed, Exists, or will Exist. However, a Fictional Entity is one from a work of fiction, such as a novel, movie, legend, etc. Within a Fictional story, an Entity could have Existed in the past of the story or currently Exists as the story is told.

SubEntities and SuperEntities

Entities are commonly subdivided into smaller parts and represent a type of hierarchy. There are numerous types of hierarchies, which are a means or showing the physical or conceptual relationship between Entities in a Group or SubEntities of an Entity in order of importance or degree of specificity of their classification [7]. For the current discussion, we are mainly interested in what are known as Compositional Hierarchies. A 3-dimensional example of this type of hierarchy is the human body, which could be represented in the followings way to show the hierarchical relationships:

- Body
    - Head
        - Eyes
            - Retina
            - Cornea
            - Lens
        - Ears
        - Nose
        - Mouth
    - Arms
        - Hands
            - Fingers
    - Legs

- Feet
    - Toes

However, Figure 1 is more representative of the actual relations between the SubEntities:

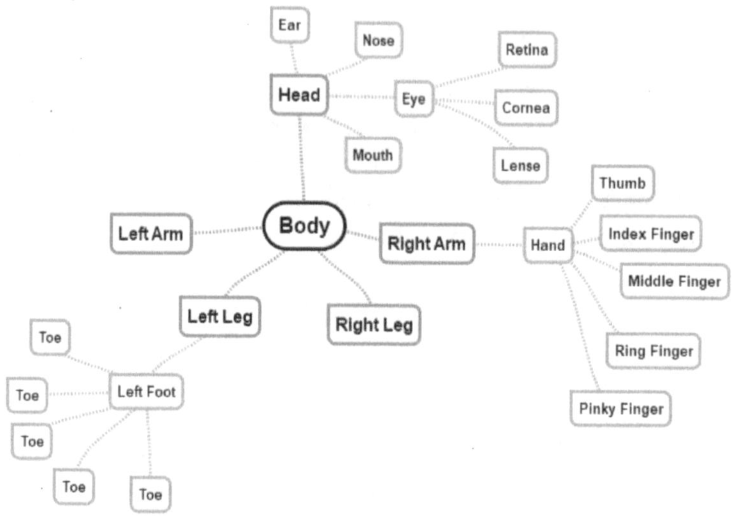

Figure 1. Relationship between SubEntities.

Head is a SubEntity of Body, Eye is a SubEntity of Head, and Retina is a SubEntity of an Eye. And it is the reverse for SuperEntities. Each of the SubEntities would have its own Attributes (this term will be described in detail shortly) to describe it to whatever level of detail is available and/or necessary. However, in addition, how a SubEntity is attached to its SuperEntity and its own SubEntities must be described and stored in memory. This represents a type of physical hierarchy.

For most Entities, we only see the surface features, and that is all that we are able to store in our memory about a given Entity. For example, if a person is observed off in the distance, only the shape of the body and possibly the main color of the clothes will be perceived. But, as I approach the person, I begin to see facial features, body build, and more details of the clothing. If I were an ophthalmologist, with the correct equipment, I could go into even more detail and look into the eyes and study the cornea and retina.

With an engine, we first see the shape, the outer surface, its color, maybe some oil in spots, but there is no indication of how it works. A mechanic, in the process of troubleshooting, might unbolt the oil pan or the camshaft cover to observe its SubParts or SubEntities, which include the crankshaft, pistons, bearings, spark plugs, and the SubEntities of those SubEntities.

With an on-line mapping application, we might start at a high level, in which we see a map of the United States with borders for the states, the major rivers, lakes, highways and cities. As we zoom in, smaller rivers, lakes, highways, and cities become visible. At the lowest level, we see streets and stores.

An example of SubEntities with varying dimensions is a terrestrial globe:

- Globe (3D (volume))
  - Surface (2D (area))
    - Oceans (2D)
    - Continents (2D)
      - Countries (2D)
        - Borders (1D (line))
        - Cities (0D (point))

A chess board is another example, in which each of the 2D squares is a SubEntity of the 2D board.

In addition to specific SubEntities, Entities and their SubEntities might be further divided. When a person looks at an Entity (or a SubEntity), he/she may consider the top, middle, bottom, back, left side, right side as being different sections with distinct Attributes (to be discussed in the next section).

Attributes

It is not known how the human brain stores the many aspects of an Entity, such as size, weight, color, temperature, etc. I would like to refer to these details as Attributes. I am not trying to describe the biological basis for the association between these aspects and a given

Entity, but rather to create a type of construct that allows us to think conceptually about an Entity. But also, I think that using the concept of Attributes could form the basis for how an AI makes the relationship between Entities and their many details or aspects.

If Entities are nouns, then Attributes are somewhat like adjectives in that they represent most of the details about an Entity and are key to enabling the comparison of Entities, especially very similar ones. I have already discussed a few types of Attributes, which include *Dimensions, State of Being,* and *Existence.* There are several different categories of Attributes, and the types of Attributes vary with the type of Entity. Often, the specific values for any given Attribute of an Entity (such as dimensions, weight, temperature, etc.) are not known, but these details can be added as they become available. For most Entities that we know about, even ones that we are very familiar with, we do not know every detail, or Attribute. But our minds create a set of memories as a mental construct of a given Entity that has as many Attributes as we are currently aware of, and we add additional Attributes to the mental construct as they become available. For example, your friend John might have told you that he's had a rescue dog for about 3 years that is very good at catching a frisbee. You are able to create a mental construct for John's dog, along with a few of its details, or Attributes, even though you may not know the dog's name, breed, age, color, or temperament. But as soon as John tells you some additional details, they are added immediately to the memories associated with his dog.

In the next sections I will discuss briefly some of the most significant Attributes that many, and possibly most, Entities have. Figure 2 shows a diagram that includes these Attributes.

<u>*Name*</u>

Unidentified Entities usually have generic names such as "rock," "dog," car," or "that guy." Known Entities that do not have proper names might be called "my car," "my neighbor's house," or "John's desk." However, Entities with proper names might also have some generic names, such as Mike, Joanna's husband, Uncle Mike, my uncle, etc.

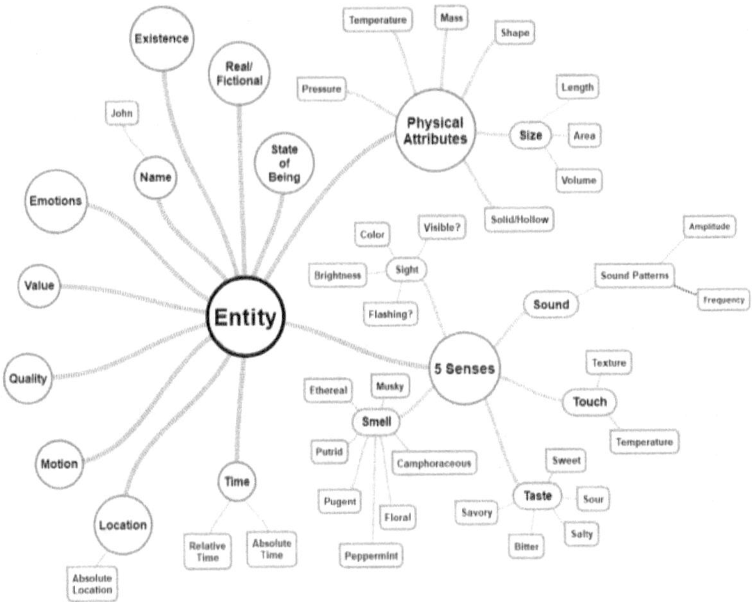

Figure 2. Relationship between a typical Entity and some of its main Attributes.

<u>*Physical Attributes*</u>

Some of the most important physical Attributes are *shape, size (3D lengths), length, area, volume, solid or hollow, mass, temperature, and pressure*. However, scientists and engineers are often interested in several other physical Attributes for substances, which include *density, specific heat, heat of vaporization, viscosity, surface tension, and thermal conductivity*. There are many other physical Attributes that could be considered, depending on the use of, or interaction with, a given Entity. An Attribute for a given Entity may not be known initially, or may not be of initial interest, but becomes important later. Therefore, it is important that Attributes and their values, of whatever type, can be added to an Entity as needed.

Often, Physical Attribute values will only be assigned a rough estimate. A person might look at a large boulder and estimate that it is so many feet high and wide but not even try to guess at the weight. If the person even thought about the temperature of the boulder, he/she would probably guess that it is the same as the air temperature, which also might be a guess.

Various methods are used to determine Physical Attributes. Guesstimation, as just discussed, measurements with rulers, balances, or other electronic equipment, information from knowledgeable people, searches on the internet or in books.

As mentioned in the section on SubEntities and SuperEntities, the value of each of these physical Attributes may vary across an Entity or SubEntity. For example, the shapes of humans and of other animals are complex even at the SubEntity level of a finger or ear. And subparts of these SubEntities, such as the top, bottom, sides, front, and back will need to be considered as to what the shape is in that area.

Humans tend to mentally simplify shapes to make them easier to remember. People will describe the shape of an Entity as being "like a sphere, circle, column, cube, square, rectangle" even if the shape is somewhat more complex. This is a way to reduce the amount of memory storage required, or even to simplify it enough to be able to remember it at all.

Attributes of the 5 Senses

All of the information about Entities is gained through the 5 senses, including most of the Physical Attributes just discussed. However, the Attributes of the 5 senses represent the information gained by the senses at a basic level, whereas the Physical Attributes require a higher level of cognition to understand them. For an example regarding sight, *bright, red, checkered pattern* would be potential Attributes. *Wide* and *cube* require mental manipulation of sight input, so would be grouped with the Physical Attributes. The main variables for each of the 5-Senses Attributes, with some details of the type of information perceived, are as follows:

- Sight
    - Visible or not
    - Color
    - Brightness
    - Flashing
- Sound

- Sound Patterns
  - Frequency
  - Amplitude
- Touch
  - Pressure
  - Temperature
  - Pain
  - Vibration
- Taste
  - Sweet
  - Sour
  - Salty
  - Bitter
  - Savory
- Smell
  - Musky
  - Ethereal
  - Putrid
  - Camphoraceous
  - Peppermint
  - Pungent
  - Floral

For sight, the colors and brightness of the visual field are interpreted in the human brain as lines of different widths and orientations, and areas. At a higher level, the lines and areas are used to perceive width, shape, or parts of Entities. Earlier, the concept was presented that the brain is designed to construct 3D images to represent real Entities. The perceived parts of Entities are inherently 2D, but when these images are superimposed on the 3D mental image, a certain part of the mental image can be filled in with the 2D perception. Once all, or at least most, of the sides of an Entity have been observed, the full details of the 3D mental image can be constructed, identified, and stored.

An AI could have the capability to perceive more details from the visual field and create a higher resolution version of a mental image of an Entity than is possible for a human brain and it could be "remembered" for a much longer period of time. And these mental machinations could be carried out much more quickly.

An AI that interacts with the world would likely have the senses of sight and hearing, and possibly touch, but may not need the ability to taste or smell. However, for certain uses of an AI, for example if it was employed in a restaurant, it might be advantageous for the AI to be able to understand that a human senses taste and smell, and how humans react to them, but it may also be desirable for the AI to have those senses itself.

For some applications, it may be advantageous for an AI to have extensions of these senses. For example, the ability to see ultraviolet and infrared light, hear ultrasound and infrasound, and to sense temperatures and pressures outside the range of human capabilities.

In addition, an AI may have additional senses, such as for radio waves, magnetic fields, and electrical voltages and currents.

Time

Time is an inherent concept that manifests itself in numerous ways. Elapsed time and time's role in motion are covered later. However, the time that a specific state of an Entity began and when it ended are two important Entity Attributes.

A state represents a snapshot of an Entity. For example, a person on a specific date and time in the past had a certain age, weight, skin markings, possible diseases, location, abilities, knowledge, etc. Depending on the context, a state might start on a specific date and time, and end a minute later. Or the time frame might be a month or more.

What has been described are absolute times, but relative time can also be an Attribute. Instead of using a specific date and time, the number of hours since some Event, or since the beginning of a certain state for an Entity, could be an Attribute.

Location

This is another important Attribute that specifies the physical location during a specific state. Location can also be absolute or relative. An absolute location on a 2D basis could be a physical address in a city or a GPS location, with altitude included for a 3D location.

Relative location is used to specify the distance between Entities. Often, distances in 2D or 3D are required. For 2D, the relative distance can be represented based on the "x" and "y" distances on cartesian coordinates, distance with cardinal directions (N, S, E, W), polar coordinates (distance and angle), etc. For 3D, the "z" axis, altitude, or an additional angle specification must be included. But algebra or trigonometry would be required to actually calculate the distance based on the "x" and "y" (and maybe "z") distances.

Relative location will be discussed in more detail in the later Chapter on Groups, Associations, and MultEvents.

Motion

This Attribute is based on a combination of location and time, since it represents the distance moved in a given amount of time. There are three main subcategories of motion:

- Stationary
- Velocity
- Acceleration

Stationary simply means that the location of an Entity is not changing as time progresses, or it has zero motion. Velocity means that the Entity is moving a specific distance for a given period of time, and the change in distance with time is not changing. For example, if an Entity moves a specific distance in 10 seconds, it will move the same distance in the next 10 seconds. With acceleration, the velocity is changing with time. If the Entity moves a specific distance in this 10-second period, it will move a different distance in the next 10-second period.

Velocity and acceleration can be on a 1D, 2D, or 3D basis and are both vector quantities, since they have both direction and magnitude. For direction, one of the coordinate systems described in the location section must be used. And motion can be absolute or relative.

Quality

Is it priceless or a piece of junk? Perfect condition has the highest quality, whereas completely destroyed has the lowest quality. Note that quality is different than value. An Entity might be a river rock that is in perfect condition (high quality), but worthless, or a 24-carat gold ring that was left on a railroad track and run over by a train (low quality) still has the inherent value of the gold.

And something might be in great condition (absolute) or good relative to something else (relative).

Value

The value of an Entity might be based on one or more of the following concepts:

- Monetary: It is worth a lot of money.
- Important Attributes: Strong, powerful, large, etc.
- Emotionally Important: Entity is loved, appreciated, needed or hated, disgusting, etc.
- Rare. This generally makes the Entity also valuable monetarily.
- Non-renewable or non-replaceable.
- Difficult to make.
- Requires a lot of material
- Requires a lot of energy
- The procedure to make it is very long and/or difficult

Emotions

People often assign emotions to certain Entities, locations, Events, words, and phrases. If a certain location evokes pleasant emotions in someone's mind, that person will tend to want to return to that location. A bad memory of someone may provoke the avoidance of that person. When writing an email, text, or letter, that is polite in nature, the author will mentally search for phrases that are agreeable rather than aggressive.

Emotions form the basis for desires, which drive the Actions of people. An AI could possibly have some form of emotions. These might be as basic as levels of good and bad. Or the AI could have secondary levels of emotions that indicate why something is good or bad. For example, if an AI was employed as a mechanic, and a certain pneumatic wrench did not work correctly several times for the AI, it would likely learn this and remember this wrench as being "bad." And, in the future, when the AI needed this type of pneumatic wrench, it would try to avoid using this one by finding another, in the same way that a human would.

Depending upon how the AI was employed, it might need to be able to understand, in some fashion, human emotions. This would probably be important for any AI working directly with people, especially in such roles as customer service or assistance. Even though the AI would likely not be able to experience the same emotions as a human, it should be able to understand how a human would react to certain situations.

Immutable and Mutable Attributes

Attributes can be immutable or mutable, in other words, unchangeable or changeable. For example, a person's name, height, shape, and weight are generally immutable (although the name can be changed, and the other Attributes listed can change slowly). Location and velocity are mutable Attributes for a person.

I consider the previous discussion as a brief introduction to Attributes. They will be covered in more detail with more depth in future sections and chapters.

Other Entity Concepts

Dimensional vs Text-based Representation of Entities

Humans mentally store Entities as 3D images and videos. However, maybe an AI would not need to store its Entity memories in the same way. So, let's consider the options. An AI database that stores the representation of Entities could do this in at least 3 ways:

1. **Text-Based**

   Words are used to describe Entities and their Attributes. Shapes would be described as "round," "square," "narrow," etc. Colors would be "red," "blue," etc. The attachment location for SubEntities on Entities would be something like "right side, 10 cm from top."

2. **3D Approximate Shapes**
   Entities would be represented mentally in 3 dimensions, but with approximate shapes, such as spheres, circles, cubes, etc. Text would be associated with the Entity and its parts, as appropriate. For example, numbers would represent the exact or approximate weight, length, area, volume, colors, etc. Colors would be understood in the AI's mind as the actual color but would have the text name associated with the part of the Entity that had that color.
3. **3D True Representations of Entities**
   This approach would mimic the way the human brain understands Entities. Brains are not perfect in how they remember actual Entities, but they try to visualize and remember them as if they were a photo or video.

I believe that the third method is the only way to achieve AGI. To fully understand how SubEntities interact with their Entity, or how Entities interact with each other, the representation of dimensional aspects is very important. If an AI with AGI-level abilities were to move a bolt into position and screw it into a nut, each of these Entities would need to be represented precisely in the mind of the AI. And if an AI were to design a new type of engine, how could it do this without an exact dimensional understanding of the Entities involved and their Attributes? (However, Text-Based representation is still used in conjunction, essentially in support of, 3D True Representation, which I will discuss in more detail shortly.)

There are several reasons why 3D true representation of Entities is important (see Figure 3):

- Identification of a known Entity that is currently being observed by comparing with the 3D memories and videos of that Entity.
- Estimation of dimensions of an observed Entity
- Estimation of proximity of an observed Entity relative to other Entities.
- Estimation of movement of an observed Entity relative to other Entities.

- Prediction of path of movement.
- Mental manipulation (using imagination) of an Entity (currently observed or not) relative to other Entities

Without detailed memories of an Entity, it would be difficult or impossible to recognize it when seen again, or to estimate its dimensions.

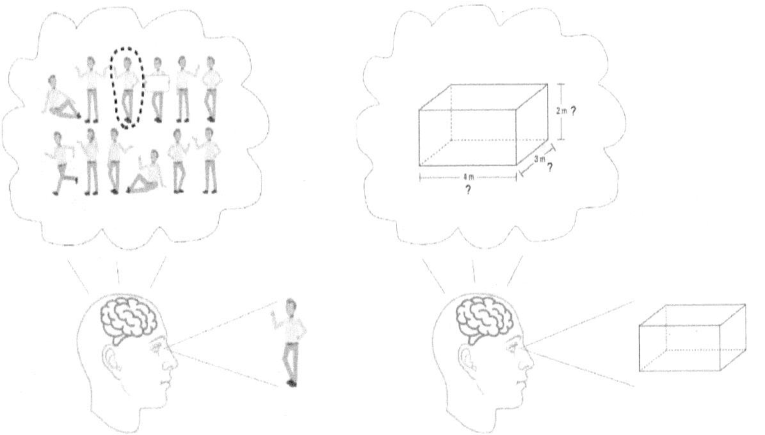

Figure 3. Identification of know Entity and Estimation of physical dimensions.

When one Entity moves toward another, two of the main things that the human brain wants to understand are (1) What is the distance between the two Entities, and (2) how fast they are moving relative to each other (see Figure 4). Another related concept is the prediction of the trajectory of a moving object based on its current movement (see Figure 5). These could probably be estimated using 3D Approximate shapes (although less precisely), but this would be difficult or impossible with Text-Based Entities.

If two Entities are to be moved together in a certain way, we humans, with our visual capabilities, understand the physical aspects of each Entity and we are able to predict or imagine how they should be moved or rotated to align as desired. To do this, we can "see" images of the Entities in our mind by retrieving them from memory, then mentally move, rotate, or change the Entity using our imagination.

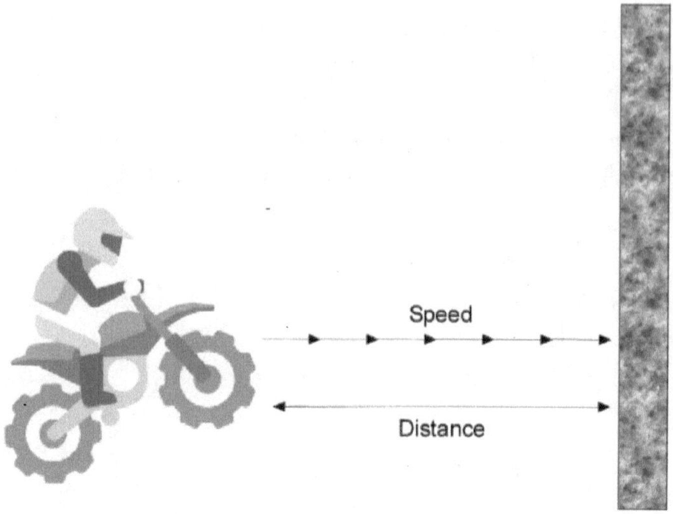

Figure 4. For Entities moving relative to each other, the human brain wants to know the speed and the distance.

Figure 5. Prediction of path for moving Entity.

But we can also see real, physical Entities and immediately use our imagination to mentally manipulate them, apparently by first storing them in memory and then retrieving them in order to mentally maneuver them. And this same mental manipulation can be accomplished by blind individuals by using their sense of touch to replace sight. All of this would be difficult to impossible without 3D True Representation of Entities.

However, the images that a human eye sees at any moment are 2-dimensional. As was discussed earlier, an innate capability of the human brain is the understanding of Entities viewed in 2D as being 3D

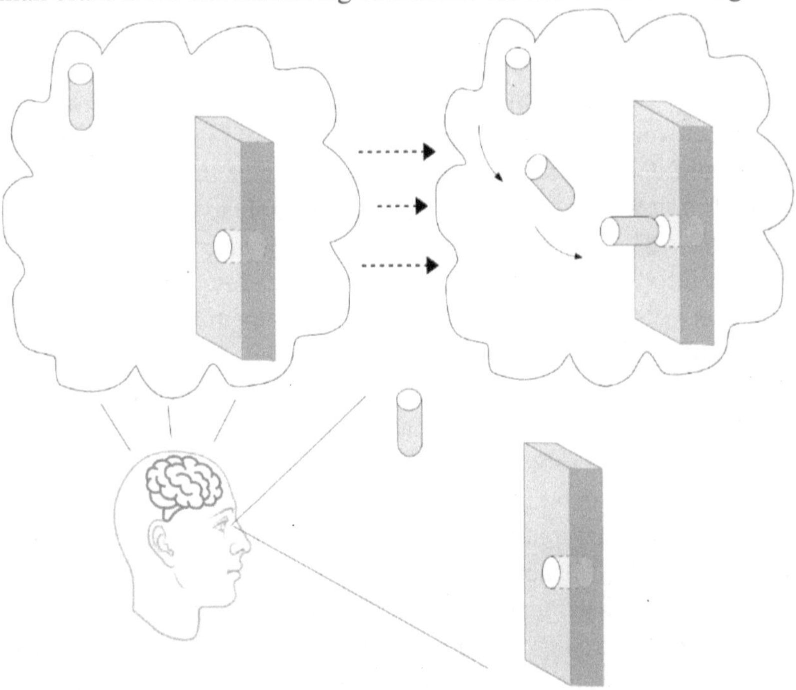

Figure 6. Visualizing how to move and rotate and Entity so that it lines up with another Entity.

objects, and vision (or touch) is used to fill in the details of the 3D object. So, by observing an Entity from different angles, the brain is able to construct 3D mental images, even though all of the visual images are in 2 dimensions. Our brain knows what the familiar Entity looks like in the back or on its side, even though we cannot see that part at the moment, or if the Entity "disappear" behind another "Entity," we can still mentally "see" any side of it. And if we close our eyes, we can mentally manipulate the Entity because of the mental construct made previously. Whether we are currently looking at the

Entities, or we are imagining them, the Entities are visualized in the same way. This would be a key capability for an AGI-level AI in order to properly understand the physical world.

If you consider the huge number of 3D, as well as 2D, images and videos that a person is able to remember, it is difficult to imagine how the human brain has this much capacity. However, according to a 2010 article in Scientific American [8], the memory capacity of the human brain is equivalent to 2.5 petabytes (or 2.5 million GB). The average movie contains about 2 GB of information and is about 2 hours long. If the human brain was a computer storage device, 2.5 petabytes would be enough to store 285 years of movies. But the human brain does not store many of the details of the images that are seen. If you watch a movie, you will not be able to remember a lot of the details. This means that, even though the movie itself may contain 2 GB of information, your brain is storing much less. This is why each time that you see a movie again, you learn something new. So, with the brain's level of storage fidelity, it should be able to, theoretically, store much more than 285 years of movies. The main point here is that the human brain has the capacity to store a lifetime of observed images and videos, both 2D and 3D.

But the human brain has additional methods to reduce the amount of information that it needs to store, and this will be discussed in next section, entitled, "Generalization."

The Text-Based method discussed above will still be utilized as part of the 3D True Representation method. Words and phrases are essentially shorthand for Entities (and Actions and Events, which will be covered in a later section), as was pointed out earlier.

Words are used for both communication and thinking. If someone says the word "horse" to you, the image and associated information are immediately retrieved. If you are thinking about a horse, the word acts as a shorthand summary of it. What you think about the horse will still be based on 3D images or mental videos of a horse, what it can do, and how it acts, but the "horse" index card can be used to call all of this up from memory.

What if we have more than one horse? Maybe we have a herd of 1000 horses. If you see this herd, you will not look at it and be able to say there are 1000 horses. The best you could do is to estimate that there are hundreds of horses. So, you will not have a mental image of

1000 individual horses. If you were to hear that in some field there are 1000 horses, your mind would probably bring up various images of individual horses and herds of various sizes. But your mind would not be able to visualize exactly 1000 individual horses. So, to record this information in your memory, you would need to use a combination of mental images of horses and the words "1000" and "horses" to comprehend what you learned and remember the exact number of these Entities (the horses).

To help understand the relationship between words and 3D mental images, let's consider that you heard that someone just bought a scrumpetflaven. Of course, this could be the name of some common item in one of the world's 7000 languages. But most of us would probably not be familiar with this word. It does not mean anything to you and your mind will not pull up any images. However, if a person tells you that a scrumpetflaven is a means of transportation, immediately images of cars, buses, trains, or maybe planes and boats, will likely begin to flash through your mind. Then if you are told it's a brand of car, you will probably first give them a funny look, since you are sure that you would have heard about this brand by this time in your life, but then visions of different brands and models of cars will likely appear in your mind.

The reason words have value and allow us to communicate and help us to think, is that they act like reference cards that trigger your brain to retrieve images and other types of information that it has previously stored. One word can cause your brain to access an Entity and all of its many Attributes and relationships to Events and other Entities. This topic will be discussed more in the chapter on Linguistics.

One other topic that should be discussed at this time is that of liquids. I believe that these are a bit harder for the brain to understand as a 3D Entity. Liquids will take on the shape of whatever container they are in, whether it is a milk jug, or a lakebed. If a glass of water is tilted to the side, the water will flow out in a stream to either your mouth, another container, the sink, the countertop, or the floor. Waves can form on the surface. And liquids can easily separate into a spray or several small puddles, and these droplets and puddles can come back together to form one large blob of liquid. The human and artificial mind has to be able to understand what is happening in these situations. This discussion is beginning to involve Actions and Event, which will be discussed in detail shortly, so I will wait to continue with this topic.

Generalization

The American Kennel Club recognizes 199 breeds of dogs, and the average person probably recognizes the most common breeds, such as the German Shepherds, Black Labrador, Golden Retriever, etc. If a person has a Black Labrador for a pet, they could probably describe in great detail the Attributes of their dog, such if it has white on the tips of two toes, the way it holds its tail, what activities it likes, how often you get licks. However, if you see a Black Labrador by the side of the road on the way into town, you may not notice or remember so many details. You will know that you saw a Black Lab, and not a Chihuahua, but probably not with the same level of detail as your own Black Lab.

If you are a breeder of Black Labradors, or have had several as pets, you would probably pay attention to specific differences, as mentioned above. However, for the average person, after seeing several Black Labs, they would likely have a mental image of these dogs that is essentially an average of the ones they had seen. In other words, the 3D image of the Black Lab is an abstraction, or a generalized version. Of course, they may remember that some of the Black Labs that they have observed had white on their chest or on their toes, but many of the Attributes would be averages, or generalizations. So, this generalized version would be an average of the range of values for each of the Attributes associated with Black Labs, such as size, weight, hair length, hair coarseness, toes, and chest, etc.

I believe that this point applies to a large percentage of the Entities of which we are aware, whether it be horses, cows, trees, mountains, buildings, automobiles, etc. You may know the difference between an oak, hickory, maple, spruce, tamarack, pine, or cedar. And you probably know the difference in common automobile brands and models. But it is unlikely that you remember the exact branch structure of every cedar tree that you have seen, or every detail of every Honda Civic that you have observed. The 3D image of each of these stored in our memory is a generalized version. So, as we humans work our way down the hierarchy of each type of Entity, we reach a level of specificity in which we will no longer remember differences between similar Entities, and we then generalize the 3D image and Attributes of that Entity.

This is a technique that the human brain uses to conserve memory space. Imagine if you tried to remember every possible detail of every tree or automobile that you had ever observed. That would be an impossible task for anyone even if they were a specialist in that area. You are likely to be able to remember several specific examples for each of these types of Entities, and almost every other type of Entities. But there is a limit, which is far below the total number in existence.

For an AI to have the same capability as a human to remember Entities, it would need to have the same abilities as have been discussed. However, it would also certainly be possible to endow an AI with much more memory capacity than a human. This could allow an AI to remember many more specific examples of any given Entity type than a human and possibly go to a deeper level of specialization for each type of Entity. For example, cedar trees with ranges of branch diameters or numbers of branches might be a further specialization of this type of tree, which the AI could differentiate and remember. With this capability, an AI could likely be superior to a human with regard to generalization.

Specific-Generic Entities

An extension of generalization is the concept of Specific and Generic Entities. If we mention "horses" or "a horse," without referring to any specific horse, this would be an example of a generic Entity. In this case, "horse" would represent the generalized concept of this type of animal, with the associated generalized Attributes. But we might be more specific and refer to Appaloosa horses. However, this would still be a reference to a generic version, but now of Appaloosa horses, which would be the average of the various Attributes associated with this breed of horse.

If we now talk about John's horse, which is an Appaloosa, we would now be referring to a Specific instance of an Appaloosa horse. John's horse would have a narrow set of the Attributes of all Appaloosas. This same concept applies to all types of Entities. Keep in mind that the Generic version of every type of Entity of which we are aware was created in our mind by seeing several, or a large number, of Specific versions of a given type of Entity, and then averaging the Attributes to form a mental Generalized version. By identifying a Specific version of this type of animal (Appaloosa horse), we are simply working with one of Entities and its Attributes that could have been used to create the Generalized version.

Self

A person has knowledge of what they themselves, or their Self, look like, their abilities, their desires, their understanding of how the world works, and their memories. In some respects, we think of ourselves as an Entity, interacting in the world much like other similar Entities (other humans). We know more about ourselves than about any other person, but many similarities exist between how we think of other humans and about our Self. For example, appearance, knowledge, abilities, beliefs.

I believe that an AI would also need to think of itself as an Entity, in much the same way that it thinks about and understands other Entities. The AI would, therefore, have stored in its memory everything it knows about John, Denise, Bob, Carmen, and Self.

Let's assume that the AI is a humanoid-like Entity. The physical Attributes of the AI would be stored as part of its knowledge database of Self. If a human were to ask the AI about its size, weight, color, etc., the AI would retrieve this information from its Self-database and report it. If the AI did something, the activity would be stored as an Event (I will discuss Events shortly) in the Self-database of the AI, in a comparable manner to the way in which it would store a similar Event for another humanoid robot, or maybe even as a human.

So, even though the AI is observing, doing, planning, and remembering in what we think of in the human brain as the mind's eye, all of the memories of these activities, learnings, etc. are stored in the Self-database in the same way that the observations of people and other Entities are stored as part of those Entities in the Entity database.

The Thoughts of Other Entities

Humans constantly try to understand what other humans are thinking. We can never have a complete understanding of what goes on in another person's mind. However, we can understand quite a bit. For example, I may very well know some of the following things about my friend Pete:

- He knows me
- He likes to talk to me
- Some of his likes and dislikes
- What he plans to do this weekend

This list could be very long, depending on how well I know Pete.

An AGI-level AI should also have the ability to know or anticipate what people and animals that it is interacting with are thinking. The AI will not experience hunger, thirst, fatigue, human desires, etc., but it should understand that a human or dog does experience these things and how it affects them. It should also be able to know what a person likes or dislikes, what it might want to do, what it might be planning.

If for example, a humanoid AI is helping a person and is asked by that person to do a task, the AI will need to make many decisions as it carries out the task. It would be important to know what the likes and dislikes of the person are so that the AI can accomplish the task as well as possible to that person's satisfaction.

Probability

This concept applies to everything that will be discussed in this entire book. But I will focus on its application to Entities at this time. Humans are always verifying everything they learn. Someone might tell you that Fred, who you have not seen in a long time, now has grey hair. If you trust this person, then you can be fairly certain that Fred really does have grey hair. But you may not be 100 % certain until you see Fred, or a photo of him, for yourself.

You may be going skiing tomorrow, and the conditions are, unfortunately, just right for it to be icy on the slopes. You are hopeful that will not be the case, but you will not be certain until you are actually skiing. Even then, if you find that all of the runs that you have tried are icy, you may still not be sure about all of the runs that you have not tried. You could make a very good guess that the other runs will also be icy, but you will not know with 100 % certainty until you actually ski the other runs, or a friend you trust, or a ski patroller, tells you about the conditions they experienced.

Someone may tell you that a certain road is closed, but you heard a news report a few minutes ago saying that the road was open. You now went from being fairly sure that the road was closed to being fairly sure that it is open.

Every Entity, its Attributes, and locations (and Actions and Events) also have probabilities associated with them. A person's belief that an Entity or location exists, or has a certain state, can vary from being sure that it does not exist or is not in a certain state, to being sure that it does exist or is a certain way.

This concept plays an important role in prediction, planning, and implementation. You need to be sure you are solving the correct problem, there are probabilities of success associated with each plan that is considered, and there is always some probability that the selected plan will not work.

Actions

Actions are about change. Change to the physical state of an Entity, its location, its motion, and for higher animals and humans, change to its knowledge and emotions.

Entities have many Attributes, which include location, knowledge, and emotions. Here, I will define the State of an Entity as referring to its physical Attributes, Attributes of the 5 senses, quality and value, all of which were discussed previously as Entity Attributes.

The state of an Entity might last many years, even thousands of years, if we are referring to a rock. However, a state might only last minutes or seconds for something being destroyed. And for a person aging, a state might last for months or even years. When we talk about an Entity changing from one state to another, we are really talking about changes in one or more Attributes, and a change is an Action.

Listed below are some of the main types of changes that an Entity can undergo:

1. Change of State (change in Entity's Attributes for Physical, 5 Senses, Quality, Value)
2. Movement of SubEntities (subparts or appendages)
3. Change of Location, Orientation, or Motion
4. Transformation (change in State of Being)
5. Relationship to other Entities
6. Emotions
7. Thinking
8. Communication

A change of state can either occur to an Entity, or an Entity can cause a change of state in another Entity. This also applies to location, orientation, and motion. We can either move or cause another Entity to move.

Movement of subparts, for a human, would refer to movement of the arms, legs, head, fingers, toes, eyes, lips, etc.

Transformation refers to a change in state of being. One example of this would be water freezing to form ice or evaporating to form steam. Another example would be a chemical reaction in which one or more types of molecules are converted to different types of molecules.

An Entity's relationship to other Entities is related to Groups and associations, which will be discussed in significant detail in a later section. However, just to mention this concept briefly here, there can be a physical change in the relationship of two or more Entities, such as them moving away from each other, or it could be more conceptual, in which there is a change to an organizational chart.

The concept of a change in emotions is rather straightforward. However, thinking and communication are more subtle in how they can be considered as a change. Thinking and communication both change one's understanding of the world, causing a change in the state of one's brain. And I think that everyone can accept the idea of thinking and communication as being Actions.

I want to briefly mention forces here. Action Types 1-4, and possibly 5 (depending on whether the relationship change is physical or organizational), require some type of force for changes to occur. Some examples of forces are gravity (force = mass x gravitational acceleration), pressure (force per unit area), temperature (temperature difference drives heat movement, which is a form of energy transfer and can be related to force x distance), and human or animal muscle force. For example, if a person steps off of a cliff, gravity causes the person to fall and is responsible for the resulting ill-effects that the person will experience. Or a ball can be set in motion by the use of human muscle force. Energy is required to make physical changes or changes in motion, but energy represents a force being applied over a distance. Forces will be discussed in more detail later.

When a person thinks about an Entity, he/she realizes, consciously or subconsciously, that it will have many Attributes associated with it. But the person will also understand, at least to some degree, that the Entity has certain types of Actions that it can do or be involved with. So, a person, a dog, or other advanced animal will be thought of as being able to do many different Actions, compared with a rock, with which very few Actions will be associated. Likewise, from the other side, a person understands that Actions have associated with them the Entities that are capable of doing that Action.

Also, subparts (or SubEntities) of an Entity will be able to do different Actions than other subparts, or the Entity, as a whole. For example, if a person is running, the legs will have one kind of Action, the arms will have another, and the person, as a whole, will be associated with Actions that are a composite of the Actions of the subparts.

An AI will need to have these same abilities to know what Actions are associated with a given type of Entity and what Entities are associated with specific Actions. However, an AI could have higher capabilities here because it could have a larger memory to store this information, allowing the AI to be able to store this information more accurately.

Now, I would like to delve into the different Action types in more detail.

Change of State

State represents physical status of an Entity between changes. For a person, typical changes to state could be an injury, aging, an increase or decrease in weight, change in appearance (haircut), or even a change in the way the person smells after an exercise session. For other types of Entities, a car for example, the Entity could change its state slowly with age or quickly from an accident. These changes are also associated with changes in value and quality (a totaled car has relatively low value and quality). Constructing an Entity represents a change in state, as SubEntities are assembled together, just as disassembling or destroying an Entity would also cause a change in state.

A change of state represents a change in the Attributes of the Entity or of its SubEntities, as well as the addition or subtraction of SubEntities. These changes could be caused by another Entity, by a macroscopic force, or by microscopic forces, such as those that cause aging or rust to occur.

However, a person, or an animal (or even a falling tree or tumbling boulder), could cause the state of another Entity to change. The person could affect another Entity by hitting, cutting, painting, lifting, or stepping on it. This is a small subset of the Actions that could be done to change the state of another Entity.

The main point here is that the same changes to state that could occur in one Entity could be caused to occur in another Entity (by the first or another Entity). Note that the Actions are very different whether the Entity is experiencing a change of state or if it is causing one.

Movement of SubEntities

Inherent to all animals is that they have SubEntities, or appendages, such as arms, legs, mouth, etc. These appendages are able to move, for the purpose of locomotion, consumption of food and water, defense, etc.

When comparing different types of animals, there will be many similarities and differences for a given type of Action. For example, consider the Action "run." For a human, dog, and insect, "run" involves movement of the legs, as well as other body parts, and the concept of forward movement. However, the human is moving two legs, the dog, four, and the insect, six. Since the legs for each of these Entities are designed differently, the specific movements of each type of leg will be different. Also, the average speed of each type of Entity will be different.

Normally, the movement of the SubEntities is caused by the Entity itself. However, in some cases one Entity can cause the movement of the SubEntities of another Entity. Examples are someone playing with their dog, a physical exam at the doctor's office, and holding hands with someone.

Change of Location, Orientation, or Motion

Change of location represents, obviously, an Entity moving from one location to another. This could be the Entity itself (a person walks to the other side of the room), or one Entity could be causing another Entity to move (a person throws a ball). The movement could be in any or all of the three dimensions. Also, it could be an absolute move, such as from this location to that location, or a relative move, such as a certain distance from the initial location, or movement relative to another Entity. The initial state would be at one location, and the final state would be at the new location, although it would probably be more accurate to say the next state, instead of the final state.

As discussed in the previous section, movement of a person from one location to another, by walking or running, involves the movement of the legs. So, this type of movement involves not only change of location, but movement of the SubEntities of an Entity. This topic will be discussed in more detail shortly.

If time is taken into account, then the motion might have a constant speed, or, if the speed is changing, the Entity would be accelerating or deaccelerating. If an Entity travels for an extended period of time at a relatively constant speed and finally stops at a new location, then we could think of the movement as three states. An initial state in the first location with no motion, the next state with a specific speed, but with no specific location, and a final state at the new location with no motion. If we account for how the speed and direction might change with time and the countless locations that might be involved, there are an unlimited number of scenarios for what represents each of the states. A human brain will try to categorize these in a way that simplifies the movement so that it can be understood and remembered. An advanced AI could potentially remember with a higher degree of accuracy the movement that actually occurred, for example, all of the locations during the movement, and the exact speed at each location and time.

Change of orientation is another type of movement. The Entity could rotate in any of the three dimensions while staying in the approximate same location or while moving from one location to another. These rotations would have speed associated with them in that the rotation would occur at some rate that could be represented as a specific RPM (revolutions per minute) or a changing RPM, on a given axis. To add more complexity to this, the Entity might be bending as it is rotating, so that one subpart is rotating differently than another subpart. An example would be a high diver that does a somersault and then straightens out before dropping into the water. The angle of the diver's body is changing, so different parts of the body are rotating at different speeds and in different directions. Similar to change of location, change of orientation can occur to an Entity or an Entity can change the orientation of another Entity (a person makes a top spin or turns a book over).

Change of location or orientation, or the associated motions or change of motion, are all Actions that are of a type specific to each type of Entity.

Transformation

Transformation refers to a change in the state of being. The main examples of this would be:

- Phase change between solid, liquid, gas
- Chemical reaction
- Nuclear reaction

For the first type, the appearance, flow properties, density, possibly temperature, and other physical properties are changing, but the molecular structure is not changing, and the phase change can be easily reversed with heating or cooling. For the second type, the molecules have been changed, along with all of their physical properties. And for the third type, even the atoms have been changed.

The main point about transformation is that it involves an Entity disappearing and a different Entity appearing in its place. For an ice cube, we have a 3D Entity that has a certain size, shape, hardness, and visual appearance (normally partially opaque). When it melts, the ice cube disappears, and liquid water appears in its place. The water does not hold its own shape and will flow to the low point it can, and it is likely clear, but definitely visible. The mass, structure of the water molecules, and, certainly, the types of atoms, will be all the same. When the water is heated and turns to steam, the water disappears and the steam replaces it, which will initially be visible, but quickly disappears and dissipates into the air. The mass of water added to the air will exactly equal the mass of water that evaporated.

A chemical reaction usually involves one molecule breaking in two, two molecules combining to one, or two molecules combining and then breaking apart into two different molecules. In any case, the original Entity or Entities are gone, replaced by one or more new Entities. The mass before and after will be the same, but most of the physical properties of the new Entities will normally be very different.

Nuclear reactions normally involve either fission, in which the atom breaks apart into a different atom or atoms, or fusion, in which two or more atoms combine. In either case, the results are similar to chemical reactions, in that the properties are much different. But now, the differences are even more profound.

Actions related to the relationship between Entities, emotions, thinking, and communication will be discussed in later sections that focus on each of these concepts. The main point to remember is that any change to an Entity, in any way, represents an Action of some type.

Complex Actions

Actions can be simple or complex, with varying levels of complexity. Moving an Entity from one location to another would probably represent a simple Action. Baking a cake would be a complex Action, since many SubActions are involved, such as grabbing a pan, mixing ingredients, heating in oven, etc.

Events

Events represent the change that occurs to an Entity (or by an Entity) due to an Action at a specific location and time. So, Actions are generic versions of specific Events. A certain type of Action describes what a certain type of Entity could do. An Event describes what an Entities has done, is doing, or will do, with specific details as to where and when the Action, or Change, occurs.

Events can involve any of the types of Action described earlier:

1. Change of State (change in Entity's Attributes for Physical, 5 Senses, Quality, Value)
2. Movement of SubEntities (subparts or appendages)
3. Change of Location, Orientation, or Motion
4. Transformation (change in State of Being)
5. Relationship with other Entities
6. Emotions
7. Thinking
8. Communication

Events encompass the first four of the five W-words, Who, What, When, Where, and Why (why refers to causal relationships, which will be discussed in a later chapter). Whenever someone wants to understand what has occurred, the values for the first four W-words usually must be determined. If only two or three of the values of these words are known, the person will usually exert some effort to learn the missing information.

The four W-words represent the minimum amount of information usually needed for the understanding of many types of Events. However, other information is often available to help better define the Event, and, often, additional details are required to fully understand the Event.

For example, the first W-word is "Who," but there is often more than one "Who." If the Event involves someone using a bat to hit a ball to another person, then we would have the subject, direct object, indirect object, and what I will call a "Tool."

When we are dealing with Events that involve movement, several scenarios are possible, and the following are generic examples of some of the main ones:

1. A person doing some Action by themself
2. A person going to another person or location
3. A person bringing an Entity to another person or location
4. A person using a tool (maybe a bat) to transfer a ball to another person or location

Grammatical terminology, such as subject, direct object, and indirect object could be used to label each of the participants in an Event. However, I would prefer to think about and label these in a slightly more practical way.

- Initiator – The Entity that is responsible for the Action. Grammatically, this would be the subject. A Location could be associated with the Initiator.
- Transient – The Entity that is transferred from one Entity to another. This could be a ball, package, pencil, cup of coffee, etc. Grammatically, this could be a direct object. The transient would go from the location of the initiator to the location of the recipient.
- Recipient – The Entity that is at the final location for the Event or is the Entity that receives the transient. Grammatically, the recipient would be a direct or indirect object. A location could be associated with the recipient, but the location could be more general and could include both the initiator and the recipient.
- Tool – The Entity that is used by the initiator to help carry out the Action for the Event.

I believe that the meanings of initiator, transient, and recipient are fairly evident. Here are a few examples of tools:

- She drove (car is the tool)
- He flew (airplane is the tool)
- She hammered (hammer is the tool)
- He baked (pans, spatula, oven are some of the tools)

The details that may be needed to describe any of the Event types can be called Event Attributes:

- Initiator

- Transient
- Recipient
- Tool
- Action
- Time
- Location
- Motion
- Frequency
- Manner
- Force involved
- Emotion

Note that some of these Attributes are mandatory and some are auxiliary. For example, only if an Entity being transferred is Transient needed. There may be no recipient or tool.

If an Entity is being moved from one Location to another over a period of time, then motion is involved. If the Event is about a rotating engine, then frequency would also be involved (RPM or revolutions per minute).

Manner refers to the way in which the Event occurred and essentially represents an adverb modifying a verb (Action). As an example, a person could drive to the store slowly, rapidly, safely, or recklessly.

Often, there is a force involved. If a person throws a ball, it is human force. If a hydraulic jack is lifting something, then the force is pressure in the hydraulic fluid. For an explosion, the force is due to a rapid temperature and pressure increase.

Emotions of some kinds are usually associated with Events. You were likely to be happy, anxious, sad, depressed, or relieved anticipating the Event or reflecting on it later. For an expected Event, it is very possible that you experienced one set of emotions anticipating the Event, another while the Event was occurring, and a different set of emotions after the Event had occurred. This topic will be discussed more in the section on Needs and Emotions.

SubEntities and SubEvents

For many Events, the Entity may have several SubEntities that are moving independently, each with their own SubEvents. Here are two examples:

As a person is walking:

- Legs are moving in a certain pattern
- Toes could be wiggling
- Heart is pumping
- Lungs are breathing
- Intestines are digesting
- Brain is thinking

As a reactor fluid is being stirred:

- Fluid is being stirred, so it is rotating and mixing
- Bubbles could be rising through the liquid
- Molecules could be rearranging

These activities would need to be accounted for either as Events of SubEntities or SubEvents of the Event. These would be hierarchical relationships since they are subparts of the Event.

Dimensional Representation of Events

Just as for Entities, the AI database that stores the representation of Events can do this in at least 3 ways:

1. Text-Based
2. 3D Approximate Shapes
3. 3D True Representations of Entities

As with Entities, the last option would be required for an AI to have human-level abilities to understand Events. If the AI were involved with high-level cognitive activity related to the real world, such as with engineering, architecture, medicine, product design, equipment repair, etc., then 3D true representations of Entities and their associated Events would be required. Some examples to consider are:

- An engineer designing an engine, with all of its moving parts, that provides the required power, starts and runs well, can be assembled and repaired efficiently, and is economic to build.
- An architect designing a building, which its electrical, plumbing, and ventilation systems that serves its needs for commercial or residential space, is structurally sound, aesthetic, safe, and economic to construct.
- A surgeon operating on a person, who must be able to repair the injury, while avoiding damage to blood vessels and nerves, and perform the operation quickly enough to avoid prolonged exposure of the patient to anesthesia.
- A product designer developing a new car, which must be aesthetic, functional, cost effective, and completed in a timely manner.
- A mechanic troubleshooting and then repairing a car, who must understand many aspects of it, including its electrical and mechanical systems, how and why they fit together, how each system is designed to operate, and what symptoms are exhibited with certain kinds of malfunctions.

Each of these activities requires detailed and precise understanding of how SubEntities fit together, operate, and affect each other.

Events are Short Videos

Whereas an Entity is represented as a 3D or 2D image, an Event can be represented as a short video of 3D or 2D Entities interacting by themselves or with other Entities. The Entities would be referenced from a central database of Entities and would need to be understood well enough as 3D images that they can represent the Actions in the Event. It could be that the Entities are known, and the current Event is another one that is added to the database of Events for these Entities. Or it might be that the first encounter with the Entities of the Event happens by observing the Event, and then the other details of the Event (W-words, for example) would be filled in, and the images for the Entities would become the first images for the new Entities which are now part of the central database.

The mental video of the Event may only last a second, or it could be quite long, depending on what the Event entails. It also may represent a large or small physical area. For example, the Event might involve two small pieces of an apparatus being carefully moved together, while slightly rotating one of them, until they touch, or a bird flying in the distance just above the treetops, or the small image of a car moving from one city to another on a map.

Human memories (mental videos) of Events are often lacking in detail. A human brain is only able to store a small amount of the information observed for an Event, and it must select the most important aspects. However, an AI would be able store much more information since it would operate much faster and would have much more storage capacity.

Mental Organization of Events

The previous section of Dimensional Representation of Events, the discussion may have appeared to imply that if you have 3D true representation of Entities, then you would not have text-based representation. However, the human brain apparently has both, and I believe that an AI should also have both. And I think that this is important with respect to the mental storage and retrieval system.

As an example, John is walking from his house to the corner of Second Avenue and Maple Street on June 15 at 11:00 AM. Here, I want to propose how an AI might store this information, which I believe has similarities to the way in which human brains do this. There should be a text-based version of the Event, which is based on the concepts presented earlier:

- John (Who) (Entity)
- Walks (What) (Action)
- June 15 at 11:00 AM (When) (Time)
- From his house to the corner of Second Avenue and Maple Street (Where) (Location)

More information may be available, such as how fast John was walking, but this information would provide the core details for the Event. Each of these details would have a direct reference to the main database for that item. For example, John is an Entity that has many

Attributes, associated Events, possible Actions that John can do, etc. When John is referenced for the current Event, all of John's many details are available to provide the needed input to the Event, including how he walks. There is likely a large database of Attributes for his house, which is also likely to have been involve in many Events, so all of this information is available when his house is reference for the current Event.

Likewise, the time stamp on this Event, now also associated with John, allows someone to quickly determine what John was doing on June 15 at 11:00 AM, and where he did it.

In this way, if someone mentions John, all of the Events associated with him can quickly be found, and we can find what John was doing at each time and where are the locations that John has been to. Or if we think about what was happening on Second Avenue and Maple Street at June 15 at 11:00 AM, we can quickly determine that John was walking there.

So, the text-based information for the Event would augment the image- and video-based 3D true representation version. The text-based version of the Event not only provides abbreviated encapsulations of the 3D details of the Event, but also provides specific details, such as the exact time and date, which might be difficult to represent by the 3D true representation version.

Specific vs Generic Events

An example of a generic Event would be several horses running across a field or cars driving on a highway. In these cases, the time would not be specific but would be "during the summer" or "on some evening." The location would probably also be generic, such as "a field or highway somewhere."

The same generic Events could become more specific if the location and/or time are known, even though the horses may still be generic and the cars on the highway are generic. But if the names of the horses are known or there is a specific car and known driver for the highway Event, then each of the Events could be very specific.

So, when an Event is observed, a memory is created of the Event. It may be generic, with only a few details, or it may be very specific that covers the who, what, where, and when in significant detail. If it is very generic initially, as more information is learned, then it can be added to the original Event memory.

For example, a woman is observed driving her car on a highway. For this Event, the Entity is "a woman", the Action is "driving," and the location is "on a highway." This Event could become more specific if the observer, or AI, learns that the woman is the same one seen earlier, or even more specific would be that the woman's name is Emily, and that she is the friend of a relative. Additional details might be the name of the highway and the fact that Emily was driving from her home to a specific store. And if it was learned that the time was 3:15 pm on June 29, this information could be added to the Event memory.

Often, we do not know all of the details when we first learn of an Event, but an initial memory is created based on what little we know. As more details are learned, these details are added to the Event memory to make it more specific.

An AI should be able to remember more of the initial details and with a higher degree of accuracy. New information gained by the AI regarding the Event would likely be learned with greater accuracy than would occur with a human.

Stories (Complex Events)

Most Events that occur are actually a series of Events. Here, I will define "MultEvent" to mean a series of Events. A MultEvent can consist of two or more Events in series, or possibly some of the Events are in parallel.

Baking a cake is actually a series of Events, or a MultEvent. As was discussed earlier under the sections on Actions, here are some of the steps involved:

1. Assemble the ingredients.
2. Grab the baking pan and the required utensils.
3. Mix the ingredients.
4. Heat the oven to desired temperature (this step was initiated at this time, so turning the oven on is in series with the previous step to mix the ingredients, but the oven takes several minutes to head up, so the heating step is effectively in parallel with the next step).

5. Pour the mixed ingredients into the baking pan.
6. When the oven is at the correct temperature, put the pan with the ingredients in the oven.
7. Bake for a specific amount of time.
8. Remove the cake and pan from the oven.
9. Turn off the oven.

Many of these Events have SubEvents or the Entities have Sub-Entities, with their own SubEvents, which would need to be accounted for. For example, "assemble the ingredients" may involve going to several different cabinets (and the refrigerator), using a hand (with its fingers) and arm to reach and grasp onto a jar or bottle, then carrying each one to a central location. So, this first Event may actually represent several SubEvents that are mostly in series.

This example shows that when a new MultEvent is mentally created, after it was experienced, read about, learned from another person, or mentally envisioned, there may be several ways to organize it. An AI should have specialized subroutines that accomplish the task of determining how to organize a new MultEvent or Event and then mentally creating it and storing the appropriate information for it.

Existence

Just as for Entities, Events can be considered to be one of the four main Existence categories:

- Exists
- Existed
- Imaginary
- Non-Existent

For example, an Event could be occurring now, so it exists. Or it could have occurred in the past, so it existed. Imaginary Events have been created in the mind and may never Exist or may Exist in the future. And non-existent Events are those that could have occurred, but did not, so they never existed. For example, "He did not drop the dish." You have to be able to envision a dish being dropped to understand this, but since the "dropping" never occurred, the dish did not hit the floor and did not break. So, the Event is non-existent.

Real-Fictional

And just as for Entities, Events can be real or fictional. A real Event is one that actually Existed, Exists, or will Exist. However, a fictional Event is one from a work of fiction, such as a novel, a movie, a legend, etc. Within the context of a fictional story, an Event could have existed in the past of the story or currently exists as the story is being told, even though the Event in the real world is only imaginary.

3

Groups, Associations, and MultEvents

"There are basically two types of people. People who accomplish things, and people who claim to have accomplished things. The first group is less crowded."

- Mark Twain

This chapter covers groups of Entities, their associations, and groups of Events, which were introduced at the end of the last chapter as MultEvents. Here, associations refer to the non-interactive relationships between Entities. Relationships that are interactive, such as comparison between Entities, transformation from one Entity to another, and models, in which one Entity represents another Entity, will be covered in a later chapter.

Groups have many of the same characteristics as Entities, but they have the additional aspect of having relationships, or associations, between the Entities that are members of the same Group. MultEvents are analogous to Groups regarding Events, in that MultEvents represent a group of Events.

I want to distinguish between Entities with SubEntities or subparts, and Groups with Entities. For Entities, the SubEntities are connected together, either directly or hierarchically. The SubEntities may also function together, such as in an engine, in which the gears mesh together, and the pistons and crankshaft move in unison.

For Groups, the relationship between the Entities may be conceptual or physical. However, even for physical relationships, the Entities are not physically connected. Instead, they are associated based on their spatial and temporal relationships to each other. These concepts will be covered in detail in the next section.

Here are some examples of types of Groups:

- Collection of Entities based on proximity, such as:
    - All of the people in a city park.
    - A flock of birds.

  - Boulders in a field.
  - Cars on a section of highway
- Organizations, such as businesses, government organizations, schools, and clubs.
- Participants in a meeting.
- Members of a family or friends of a person.
- Possessions of a person, which could be all of the person's possessions, the clothes currently worn, or the items held in the person's hand.
- Classifications, such as:
  - Hierarchy of animals, with SubGroups of vertebrates, of mammals, and or of Breeds of dogs.
  - List of all of the books I have read.
- Fluids
  - Solid particles suspended in liquid or air.
  - Bubbles suspended in a liquid.
  - Droplets suspended in air.
- Chemistry
  - Zillions of atoms and molecules in a substance.

As shown, there are many ways that Entities can be associated in Groups, both physically and conceptually. And the size of Groups can vary from two to near infinity. Generally, although not always, the members of the Group will have some things in common, such as all human beings, birds, members of the same family, bubbles, etc.

And each of the Entities that are members of the Group will have their own Attributes. In the case of raindrops, we probably will not be interested in the Attributes of each raindrop, although a meteorologist may want to know the average drop size and temperature. However, for employees in a company, the many Attributes of each of these Entities are important, in varying degrees, to the other members of the organization.

The following are some general concepts that pertain to Groups:

- Members:
  - What are the main characteristics or Attributes of the members?
- Order of the Members:

   - This can be both conceptual and physical, and could refer to the hierarchy in an organization or classification system, or it could mean the physical order of the members relative to each other. In this case, order would refer to the physical arrangement, which could be in 1, 2, or 3 dimensions, but order would not indicate the location, orientation, or motion of the members.
- Location, orientation, and motion of the members, or Entities relative to each other.
- Temporal
   - A Group could be based on one Entity that changes with time, so the members of this Group could be the same Entity at different times (e.g.- a person at different ages).
- Hierarchy
   - What are the Group Entities' relative importance, or how are they categorized relative to each other?

Groups also have Attributes, which are analogous to the Attributes for Entities, and can represent some of the same concepts as Entities, but their Attributes will often also refer to different types of concepts. For example, some of the Attributes of a Group might be Group type, whether it is conceptual or physical, how many members, member type, Group age, etc. And Groups can be part of Events and MultEvents, as can Entities.

As for Entities, MultEvents for Groups are essentially a group of Events that occur in series, such as for a recipe, but can occur in parallel, such as cars on an assembly line in which the partially assembled vehicle reaches a station in which four workers simultaneously install the four tires.

I would now like to discuss Groups and MultEvents in more detail and compare the similarities and differences with Entities and Events, respectively.

Groups

With Entities, we are interested in what they are and what they do. In other words, what are the characteristics of an Entity, or what are its Attributes, how many dimensions does it have, when it existed, what is its state of being, etc. Also, what Actions can the Entity do or what has it done, or what Events has it been part of? With Groups, we are concerned with some of the same points, such as the Group's Attributes and when it has existed, but we are also interested in concepts like the Entities that make up the Group and what do they have in common, how they are associated with each other, and what Actions can they do as a Group.

There are really three main ideas that define a Group: what are the members (Entities), how many members are there, and how are the members organized, or what their order is. A sub-category for organization is that it can be conceptual or physical.

Members of a Group can be divided into at least three categories, which are random, members with certain Attributes, or specific members. Random would mean that the members may or may not have anything in common. For the second category, they would have one or more things (Attributes) in common. For example, the members might all be human, but every other Attribute might be random, or the members might be all of the cars made from the same assembly line today. For the last category, the members are specified, such as the letters of the alphabet, or the employees at a company.

The number of members can be random or specified. So, the number could be random, or any amount, or a specific number of members.

The order of the members can be of many types of organization, but the first consideration is whether the order is based on conceptual or physical organization. However, for each of the categories discussed here, the members could be ordered or random.

For conceptual organization, there can still be spatial concepts involved. The alphabet and integers are examples of 1D (one dimension) order. The letters (Entities) have a specific order, which is linear (1D). They could be physical letters on a page or 3D blocks, as for a sign, in which they would have specific physical sizes and inter-letter distances. However, the general concept of the alphabet is represented only by the basic shape and pattern of each letter and the order of the letters relative to each other, in which case the general concept is conceptual rather than physical. Integers are similar to the alphabet in these aspects.

A flowchart for a process, a decision tree, or a schematic diagram are all examples of conceptual 2D order. The blocks (Entities) representing the steps, decision points, or physical equipment can be of various shapes, sizes, and colors, although some of them might be based on standardized configurations. Also, the spacing between each of the Entities is usually not important. The most important concept is how the blocks are conceptually connected to each other, which is usually by lines that sometimes have arrows representing the direction of flow or order of progression through the diagram. Note that the connections are indicating the interaction between the Entities, as compared with the physical connection between SubEntities of an Entity.

Some examples of 3D order are 3D models of the solar system and of molecules. Often, they are to scale, regarding the size of the planets and sun, relative to each other, but may not be. The relative distances usually are not accurate. The most important point is to show the order of the planets. For molecular models, the main objective is to show how many atoms of each type and how they are connected to each other, regarding their order. The physical representation of each atom, the relative size, the distance between the atoms, and the actual chemical bonds (they are often wooden sticks rather than orbiting Entities representing electrons) are not realistic, so these models are essentially conceptual.

We can also have order based on temporal concepts. The days of the week, months of the year, the seasons, and the day-night cycle are examples of conceptual temporal order.

The concept of hierarchy [7] was briefly discussed earlier in reference to Entities and SubEntities. Some of the main types of hierarchies that are conceptual and are associated with Groups involve the relative importance, value, or power (or authority) of members of the Group, or categories describing members that vary from general to specific.

An example of importance would be the alarms for the control system of a chemical process. For value, examples are evidence for a trial or data from research, or the bills or notes for a monetary system. The management structures of governments and corporations are examples of hierarchies based on power or authority. Biological taxonomy is an example of categories that progress from specific to general.

Now let's consider order based on physical organization. These relationships can be based on space, time, or motion (which involves space and time). Spatial relationships can involve the location or orientation of Entities relative to each other, and these relationships can be based on one, two, or three dimensions.

The order of letters in a word on a page or the mileposts on a highway are examples of physical 1D order. The letters are in a specific order and are a certain distance from each other on the page. Mileposts are also in specific order and are exactly 1 mile apart. Examples of physical 2D order are a specific chessboard, the blocks in a city, and an assembled jigsaw puzzle.

The furniture, pictures, rugs, window shades, and decorations in a room are examples of physical 3D order. The models of the solar system and molecules were described earlier as examples of conceptual 3D order. However, if we consider the physical parts that are assembled to make the model, then these models could also be considered as having physical 3D order as an Entity with SubEntities, since the subparts are physically connected to each other. An engine would also be thought of as an Entity with SubEntities rather than a Group with Entities that are next to each other. This is because the SubEntities are physically connected and work together as a single Entity, similar to the body parts of a complex organism.

The location of a person at various times of the day could be considered as physical temporal order if we consider the Group as the same person at several times of the day. Another example would be the appearance (Attributes) and state of a person at different times in their life as they age. In both cases, a photo or set of photos, which would be a type of Group, could represent each state. The same concept could apply to a physical Group, in which the state of the Group vs time could be observed.

We can also consider the motion of Entities in the Group. Bike riders in a race or the individual geese in an airborne flock will move relative to each other. So, the velocity of the members relative to each other could be determined.

As was mentioned earlier, for each of these categories, the members, number of members, or order (and its sub-categories), could be random rather than having set values or a specific order.

The following outline summarizes the concepts just discussed:

- Members (Entities)
- Number of Members
- Order
    - Conceptual
        - Spatial
            - 1D
            - 2D
            - 3D
        - Temporal
        - Hierarchical
    - Physical
        - Spatial
            - 1D
            - 2D
            - 3D
        - Temporal
        - Motion

State of Being, Existence, and Real-Fictional

Some of the concepts for Entities apply to Groups, but some are very different. State of being applies when the Group is represented physically, such as liquid droplets in a gas, bubbles in a liquid, solid particles in a gas or liquid, a model of the solar system or of a molecule (when the physical nature of the parts of the model are considered rather than what the model represents conceptually).

Existence for Groups would be very similar to Entities. For example, a real Group might currently exist, or it might have existed in the past. A Group could be imagined as a possibility, maybe as a variation of an existing Group or one that did exist, or it could be a completely new idea for a Group. And it is possible to consider a non-existent Group, as the following statement indicates: "No Group wanted to be involved." You have to imagine the idea of a Group that could have been involved, but with the realization the imagined Group (or any Group) was not involved.

A fictional Group would be like a fictional Entity, in that it never existed in reality, but only within a fictional story. It may exist at some point in the story, or it may have existed in the past of the story.

SubGroups

Just as there can be SubEntities within Entities, there can be Sub-Groups withing Groups. Here are some examples:

- Divisions within a corporation
- A family unit at a family reunion
- Math books at a library
- The bad apples in a basket of apples
- The calves in a herd of cattle
- The integers between 20 and 30
- The English letters between F and M
- The cars in a parking lot that are red

SubGroups can also be part of a hierarchy, such as with a corporation or with animal taxonomy.

Group Attributes

Groups have Attributes, just as Entities, with a lot of similarities but also with many differences. These next sections tend to focus on the types of Attributes that are different for Groups, when compared with Entities.

Physical Attributes

Physical Attributes would not apply to conceptual Groups, but some would apply to physical Groups. For example, a school of fish would have a size, a shape at a specific moment in time, a volume, a mass, and a temperature. But, in general, Groups would not have properties such as viscosity, surface tension, or thermal conductivity.

Attributes of the 5 Senses

You can imagine the sight and sound of a flock of geese flying south for the winter, and the smell of a bouquet of flowers. However, touch and taste are usually associated with Entities, rather than Groups.

Time, Location, and Motion

These types of Attributes would apply to physical Groups. Time would also be applicable for conceptual Groups, with regard to the state of a Group, or when a given state began and ended for a Group. However, location and motion would not apply to conceptual Groups.

Quality and Value

These Attributes would apply to some physical Groups, such as a corporation. A flock of wild geese could be considered to be of high quality if they were all healthy, or of lower quality if some of the birds were sickly. However, value as an Attribute would not apply (but might apply for a rancher (or gooseherd) that owns a flock of domesticated geese). Neither quality nor value would apply for most, if not all, conceptual Groups.

Emotions

Just as specific Entities can elicit emotions of different types in people, Groups can do the same. You might have good feelings about the corporation (Group) that makes your cell phone, or bad feelings about the company that provides your cellular phone service.

Depending upon how an AI with AGI is programmed to understand and "feel" emotions, this AI might experience something similar to a human with regard to an organization or Group. For example, if the AI requested information, or purchased a product, or had services performed, and the information was wrong, or the product or service was inadequate, the AI would likely store the difference between expectations and results. When a similar need occurred in the future, the AI would consider this information and likely decide on a different supplier. So, the AI, in some fashion, would likely express a form of dissatisfaction, which is one type of emotion.

Emotions related to AI Entities will be discussed in more detail later, at least some types of emotions can likely be programmed into an AI. If we consider that one of the main purposes of emotions in humans is to encourage or discourage certain situations or Actions, artificial emotions could be programmed into an AI to accomplish the same purpose.

Probability

If you have not actually seen a specific Group, there is the real possibility that it is not the same as what has been described. In a large or complex Group, it may be difficult or impossible to know every detail about it. For any information received about unknown aspects, there is some probability that the information is not correct. And for the details that are known, some probability exists that it will change tomorrow, in an hour, or in a second to some state that may be difficult to predict.

MultEvents

MultEvents are analogous to Groups in that they represent a group of Events organized in a certain way, rather than a group of Entities. MultEvents can be associated with Entities or Groups. The Events of a MultEvent normally occur in a serial fashion, with one Event following another in a temporal manner. However, the Events can run in parallel with some part of the MultEvent. And MultEvents have Attributes similar to those of Events.

The following are some examples of MultEvents:

- A procedure or recipe.
- Construction of some Entity.
- Driving to a location.
- Going to the store to buy groceries.
- The progress in a meeting.
- Moving to a new location.

Almost everything that we do represents an Event that is part of a MultEvent. Similar to the relationship between Groups and Entities, MultEvents may consist of two Events or an unlimited number. It depends on what the MultEvent involves and how it is organized and related to other MultEvents. Your life could be considered one massive MultEvent with sub-levels of MultEvents that form a complex hierarchy.

As with Groups, there are three main concepts for MultEvents:

- What are the members (Events in this case)?

- How many members are there in the MultEvent?
- How are the members organized, or what is their order?

In most cases, the Events would not be random, but there could be some instances in which they are. Auto accidents consist of a series of Events that are essentially random. However, most MultEvents would represent a set of Events that are specific or belonging to a certain category. For example, baking a cake is a MultEvent that is based on a recipe or procedure. The Events can be varied somewhat but by a large amount, or the cake will not turn out correctly. If you are driving to town, you can use a different car or possibly travel on different roads, but the number of variations is somewhat limited, or you will not achieve your goal of driving to town, or at least not the town that you planned on.

The number of Events is important, but it is defined by which Events are part of the MultEvent.

The order of Events is usually serial, but some may be parallel. If you think about the activities (Events) for one of the many tasks in a given day, such as baking a cake, driving somewhere, or maintaining your bicycle, most of the Events are in series, with a few in parallel. For example, while the cake is baking, you might also be making the icing. Maybe driving is a better example. There are the main Events connected in series, but a person is constantly doing several things at once (in parallel), such as looking forward and to the sides, looking at the dashboard to see the speed and fuel level, glancing at the sandwich you're about to take a bite of, or texting, if you don't care whether or not you make it to your destination.

The order of the Events often cannot vary significantly, but sometimes it can. Driving to town will have a fairly specific set of Events that must be followed in a certain order. Assembling a jigsaw puzzle can be much more flexible or random. If you think of an Event as an activity in which a puzzle piece is picked up, then placed in the correct location, you can work in this corner for a while, then the opposite corner or the middle, selecting each of the options at will.

Just to take a sneak peek ahead at Imagination, contemplating how to modify a certain MultEvent to achieve a different goal would be associated with creativity. Using your imagination, you can start with a known MultEvent such as baking a specific kind of cake using a certain recipe. If you want to modify how the cake turns out, or you want to bake a different kind of cake, you can use your imagination to modify Events, insert one or more new Event in the sequence, or remove an Event. So, in a way, a MultEvent can be thought of as a set of blocks that are lined up and can be replaced, rearranged, or even modified.

Let's look at the cake recipe in more detail. At a high level, this involves gathering the ingredients, mixing the ingredients in a pan, and baking the cake. It appears that we have three Events that make up the MultEvent. However, each of these Events can be further divided into SubEvents as follows:

- Gather ingredients
    - Get milk
    - Get eggs
    - Get flour
    - Get butter
    - Get baking powder
- Mix ingredients
    - Get a pan
    - Get the mixer
    - Place ingredients in the pan
    - Mix ingredients with mixer
- Bake
    - Set the oven at 350 °F
    - Place pan in oven for 35 minutes
    - Remove pan from oven
    - Turn off oven

Each of these SubEvents could be divided into more basic SubEvents. For example, "Get milk" consists of:

- Gather ingredients
    - Get milk
        - Walk to the refrigerator
        - Open the refrigerator door
        - Reach for the container of milk
        - Grasp the contain and lift it up
        - Pull the milk container out of the refrigerator
        - Walk to where the cake is being made
        - Set the milk on the counter

As you can see, there is a hierarchy of Events (or SubEvents) that can go to more basic levels. "Walk to the refrigerator" may seem to be at a very basis level, but it still involves activities that are performed by the legs in order to walk. However, I think that we can consider

"Walk" to be a base Action that one would not normally consider dividing into even more detailed SubActions. I believe that for most people, when they hear or think of the word "Walk", the following concepts come to mind:

- Involves movement of the legs and lower body, of an Entity, with specific motions
- Serves to move the Entity, usually in the forward direction

Other base Actions are Run, Stand, Lay Down, and Throw. Each of these Actions has a set of concepts that define it. And when an Entity, such as a human, performs one of these, it is considered as a base Event.

So, "Baking a Cake" involves a series of many base Events, although some may be carried out in parallel. Building up from the bottom in our hierarchy, several base Events can be grouped into one concept, such as "Get milk", and then at a higher level, such as "Gather ingredients".

Your life is actually a series of millions of base Events. Depending upon how we are thinking of our past, we tend to group these base Events at different levels. For example, if you are remembering how you fixed the toaster, you will think about unplugging the toaster, getting the tools, opening the side pane on the toaster, etc. At the other extreme, you might group the Events as going to elementary school, middle school, high school, then attending college, getting married, getting your first job, etc. This last example has many hierarchical layers, and the Events can be grouped in almost unlimited ways. For example, if you had an apartment while attending college, the "fixing the toaster" MultEvent might have been at the deep bottom of a many-layered "attending college" MultEvent hierarchy (because you couldn't afford to buy a new one) along with every other Event that occurred that day. Or the first week-long vacation that you took after getting your first job could be understood as a MultEvent with many sub-levels of MultEvents and Events organized in hierarchies under it.

People might have different concepts of what the base Events are, and therefore, they would likely group the Events differently, both as MultEvents and regarding their hierarchy.

MultEvents may represent cycles, such as day-night, the seasons, the motion of a piston in an engine, someone pedaling a bicycle, the daily routine of traveling to and from a workplace. For the bicycle, this could be considered as having two base Events in which the riders left leg presses down the pedal followed by the right leg doing the same, then these two Events repeat. For a standard 4-stroke engine, such as an automobile engine, the air-fuel intake valve is open as the piston moves down, then is closed as the piston moves up, the spark plug fires, the piston moves down, and the exhaust valve is open as the piston moves up. Then the cycle repeats itself indefinitely. Each of the Events for the engine, such as the valves opening and closing, and the spark plug firing could be considered as Events that follow each other to form a cyclical MultEvent.

In addition to serial, parallel, hierarchical, and cyclical order for MultEvents, there can be converging or diverging order. With converging order, two or more Events proceed an Event, whereas diverging order implies that one event precedes two or more other Events.

Many of the Attributes associated with MultEvents are those that are part of the individual Events. However, MultEvents can have several types of Attributes, such as the following:

- Events (Members)
    - Which Events
    - Order of Events
- Time (beginning and end of MultEvent)
- Location
- Frequency (how often does this MultEvent occur)
- Emotion (feelings associated with the MultEvent)

4

Interactions

"No man is happy but by comparison"

- Thomas Shawell

The last chapter covered the organization of Entities as members of Groups, and Events as members of MultEvents. For Groups and MultEvents, the concepts of which members, how many members, and their conceptual or physical order were discussed. The association of the members was covered, showing how they might be next to each other or follow each other, but not how the members interact with each other.

In this section, four types of interactions of Group and MultEvent members will be discussed:

- Comparison
- Model
- Transformation
- Opposites

Comparison

Comparison deals with the various ways in which one member of a Group or MultEvent might be greater than, less than, or equal to another member. The comparison can be qualitative, or it might be quantitative, in that one member can be greater than or less than by a specific amount. Let's focus on Groups first.

This concept applies mainly to the Attributes of Entities. Many of an Entity's Attributes can be used for comparison. For example, physical Attributes are commonly used for comparison: size, length, area, volume, mass, temperature, pressure, etc. One Entity might be heavier than another Entity, or it might be heavier by 25 kg. Or the first Entity might simply be shorter than the second Entity, or shorter by a specific amount.

Other types of Attributes can be compared as well. When comparing two Entities, one of these might be smellier or faster or have higher quality or have a higher value.

Two Entities would not necessarily have to be members of the same Group to make and keep track of comparisons. One Entity might "know" that it is wider than a second Entity and is narrower than a third Entity. And the second and third Entities might "know" their widths relative to each of the others.

But let's say that there are many individual Entities, and we would like to order them based on how narrow they are. It would be difficult, although possible, to do this by looking at all of the paired comparisons. This would be similar to determining the overall order of all 26 letters in the English alphabet by only knowing for each individual letter what letter was before and after it. In the case of the alphabet, the 26 letters are members of a specific Group, and the Group keeps track of the overall order. Likewise, the Group concept is better suited to keep track of how narrow each of the members are. Therefore, if the Entities being compared are not already members of a Group, then a Group should be created for those members specifically for comparing and keeping track of their narrowness and any other Attributes that need to be compared between them. Then, the Entities should keep track of which Groups they are members. In this way, comparison could be made in more than one way. The individual Entities could be directly compared based on specific Attributes (such as narrowness) or the Group could be queried for a comparison of any given Attribute between two or all of its members.

This is not to say that a Group should "know" all of the possible comparisons between all of its members for every comparable Attribute. But when a query is made for this information, and the comparison is determined, then the Group should store the results. However, who or what is making these comparisons? This will be covered in a later chapter on agents. What is important to remember at this point is that two or more Entities can be compared based on many different Attributes, these comparisons can also be used to order two or more Entities based on a given Attribute, and the results of these comparisons can be stored in memory.

Regarding the concept of ordering Entities based on comparisons of an Attribute, it is interesting to compare this idea with the order of Entities such as the letters of the alphabet. For the alphabet, there is no reason that the letters are in the order that they are. As for the alphabets of other languages, at some point in the development of

English, and after the letters had been created, a decision was made on their order. You cannot logically determine what the order is or should be. You just have to memorize it.

But for Entities such as sticks, they can easily be placed in order based on their individual lengths, weights, or other Attributes. For the letters of the alphabet, the order has to be memorized, but for the Attributes of Entities, comparison can be used to logically determine the order.

An AI should have an advantage over a human regarding comparison, in that it should be able to compare more Attributes more accurately and more quickly.

Model

Model deals with the various ways in which one Entity or Group represents another Entity or Group. Here are some of the types of models:

- An Entity or system is too large to be fully observed or comprehended. A miniaturized version is created to allow a person to understand it. Here are some examples of models based on this concept:
    - Blueprint
    - Physical model of a chemical plant
    - Road map
    - Globe of the Earth
    - Orrery of the Solar System
    - Drawing of the Milky Way Galaxy
- An Entity or system is too small to be observed by the naked eye of a person. Some examples of these models:
    - 3D model of an atom
    - Drawing of a pharmaceutical chemical
    - Drawing of a living cell
    - 3D model of a microscopic organism
- A model to show how an Entity or Group is organized.
    - Parts diagram for a piece of equipment
    - Blueprint
    - Schematic diagram
    - Diagrams showing the systems (muscle, skeletal, nervous) of a human.

- A model to show how an Entity or Group physically functions.
    - 3D see-through working model of an engine
    - Functioning miniature volcano
    - Video showing how something works
- A model used to develop a full-scale Entity
    - Miniature airplane in a wind tunnel
    - Pilot plant for collecting design data
    - Prototype
- A reduced-scale replica for collectors
    - Car, boat, train, airplane, building.

So, models are used to help visualize, understand, or represent Entities and Groups that are very large, very small, are complex assemblies of SubEntities, or are functioning Entities. The real Entity and the model are conceptually paired, meaning that the model specifically represents a generic version of a certain type of real Entity, or possibly a specific Entity. And the model may be an almost perfect scaled replica of the original Entity, or it may be very conceptual, as in the case of a schematic diagram.

For an AI to learn about and understand the world in a manner similar to humans, it is very likely that models would be as useful to an AI as they are to a human. The difference might be that the AI is able to understand the connection between the model and the real Entity in more detail and more quickly than a human.

Transformation

It is a very common phenomenon that Entities will convert to a different Entity. However, there are four main ways this can occur:

- Two or more Entities combine to form a single Entity.
- An Entity breaks apart to form two or more Entities.
- An Entity converts to a different Entity.
- One or more Entities exchange SubEntities with one or more other Entities.

Some of these transformations might represent an assembly of parts such that the SubEntities still maintain their original Attributes, whereas for other transformations, some or all of the Attributes are completely changed.

For example, for the assembly of an engine, numerous parts (Sub-Entities) are bolted together to create a new Entity that has functionality that did not exist before. However, each of the parts has maintained its original characteristics and the engine can be disassembled to recover each of the parts intact in its original form with its original Attributes.

A sculptor will chip away at a large block of granite to form a sculpture, which will have a much different appearance than the block of stone. However, the color and density of the finished product may be similar to the starting material. In this type of transformation, only some of the Attributes have changed, such as the shape, size, and mass.

Ice melting to form water is an example of a more profound transformation. We now have a liquid instead of a solid, and both the appearance and the density have changed. However, the mass of the water is the same as the ice that melted, and both the ice and water are composed of water molecules. So, the appearance has changed significantly, but much is the same.

An esterification is a type of reaction in which an organic acid and an alcohol react to form an ester and water. Essentially, the alcohol molecule is joining the acid and spitting out water in the process to produce the ester. The properties of all of these molecules are very different. This includes the boiling and melting points, density, heat capacity, thermal conductivity, viscosity, heat of vaporization, toxicity, flammability, and reactivity. The number of carbon, oxygen, and hydrogen atoms in each new molecule is different from the original molecules. But the total mass of the reaction mixture is the same, and the total number of each type of atom in both reactants is the same as the total in both products. And the basic structure, or organization of the atoms within the molecules is still partially intact. So, this is a more profound change than for the previous examples, and by most indications is a complete change in Attributes, but, in reality, some Attributes have not changed.

The conversion of hydrogen to helium that occurs in stars is an even more complete transformation, in that even the atoms are changed. However, even for this transformation, two of the four protons and two of the four electrons involved in the reaction remain the same.

Here are a few other examples of transformations with varying degrees of Attribute change:

- Melting down a block of metal and creating an equipment part using a mold.

- Volcanic eruption.
- Using explosives to blow up a building.
- Using glue to reassemble a broken vase.
- Burning newspapers, firewood, coal, oil, natural gas.
- Aging person.

Opposites

Opposites refer to the following concepts:

- 1, 0
- On, Off
- Yes, No

This type of interaction is not quite like the previous ones, in that it does not involve two different Entities. Rather, it pertains to one Entity having two different states, most likely expressed by one if its Attributes. However, it is somewhat similar to transformation, but instead of an Entity changing into a different Entity, the Entity is changing to a different state.

This concept is actually very broad, and here are some additional examples:

- Front, back
- Left side, right side
- Top, bottom
- Up, down
- Forward, reverse
- Black, White
- Correct, Incorrect
- Light, dark
- Loud, soft or silent
- Tall, short
- Big, small
- Day, night
- Start, stop
- Full, empty

The list could go on and on, and applies to Groups and Events, in addition to Entities.

5

Logic, Reasoning, and Imagination

"Logic will get you from A to B. Imagination will take you everywhere."

- Albert Einstein

The subjects of logic and reasoning are extremely complex and have been covered in countless books, journals, and other publications. However, the main purpose of logic and reasoning is rather simple and involves the following: *making known what is unknown using what is already known.*

To simplify the dialog somewhat, I will use the term "logic" at times to refer to both logic and reasoning. An extremely simple example of the use of logic would be if I had 4 apples and 4 oranges in a basket and I want to know how many pieces of fruit are in the basket. I could, of course, count every piece to get 8. But if I already know there are 4 apples and 4 oranges, I can use a form of logic called arithmetic to obtain the answer. The addition operator allows me to determine an unknown value, the sum, from two known values, the numbers of apples and oranges.

This is a bit of a silly example, but it demonstrates the concept of determining unknown information from known information. Forms of logic can vary from the very simple to the extremely complex, but it is always employed for this same purpose.

Although logic has been subdivided into countless sub-topics, I believe that almost all logical thought is based on a few basic forms of logic. Imagination is not actually a form of logic, but it is so intimately involved with logical thought that I am including it here. The following are these basic forms of logic:

- Memory
- Analogy
- Causation
- Patterns
- Mathematics
- Imagination

These topics will be covered in detail in the next sections, but these sections are not meant to be exhaustive portrayals of these topics. They are meant to describe, at a very basic level, how an AI would use each of these concepts. During implementation of an actual AI based on the ideas presented in this book, it would be necessary to expand on the details for each of the forms of logic, reasoning, and imagination that will be presented in the next sections.

Memory

It seems strange to include memory in a discussion about logic and reasoning. However, it forms the basis of logic, reasoning, and imagination. For analogy, we need to have a basis for comparison of one Event or Entity with another to determine if the primary Attributes or SubEntities of both are similar enough to say that there is an analogy. To form rules for causation, we need to be able to remember the data from several Entities or Events (maybe experiments). For imagination, we may start with a known recipe and mentally change one or more of the Events to get a different, but desired, outcome. Or we may start with a known Entity, which its information contained in our memory, and use our imagination to change Attributes or SubEntities to obtain an improved or different Entity.

Our memory allows us, in a way, to predict the future. If we drive our car down the road that goes from town to our house, park the car in our driveway and enter the house, we can predict numerous things along the way. We can predict how the door latch will operate, where the ignition switch is, how the car will respond when we step on the accelerator, and how to turn on the radio. We can predict that if we follow the road the same way we did last time, that it will lead to our house, and about how long it will take. When we enter our house, we can predict where to find the kitchen so that we can put the groceries away.

If we buy a new car and new house, we will use analogy to know how to drive the new car, using our old car and possibly other cars we have driven or seen, as a basis to make the analogy to the new car about how it operates. For the new house, we might use imagination and causation/counterfactual to find the kitchen in the new house (if I had turned left instead of right maybe that would lead to the kitchen), and analogy to know which appliance is the refrigerator. But we would be using our old house, and possibly other houses, as a basis for each of these forms of logic (that new appliance looks like my old refrigerator).

The importance of memory for logic, reasoning, and imagination will become more evident in the following sections.

Analogy

Analogy may be the most important form of logic and is involved with almost all forms of thought. The essence of analogy is that when two "things" are compared, there will be certain characteristics that are of primary importance and other characteristics that are less important. The two "things" would be considered to be analogous if the primary characteristics were similar. I use the word "things" here because analogy can apply to Entities, Groups, Actions, Events, MultEvents, and other concepts that have yet to be discussed.

Most people understand the concept of analogy, even if they do not consciously think about it. But let's consider some examples to drill down into the underlying points. We recognize other human Entities by analogy. Everyone is a little bit different regarding their facial features, hair and skin color, height, weight, posture, eyewear, jewelry, and clothing. However, humans have a head, neck, chest and abdomen, two arms and hands, two legs and feet, two eyes, two ears, a nose, a mouth, hair on the upper part of their head (usually), and they are within certain height and weight ranges. This set of characteristics are the primary Attributes that we would use to identify the Entity as a fellow human. There can be some variation in the Attributes of the human body, and the person will still be identified as a human. The person's eyewear, jewelry, and type and color of the clothes are really of no importance in this identification.

When you rent a car or buy a new car, you are able to drive it with little difficulty, even though these models might be very different than the car that you currently own. Why are we able to do this? Because there are certain features, SubEntities, and Attributes that are part every car, even if there is some variation in these SubEntities and Attributes. Every car has four wheels, a steering wheel, accelerator pedal, brake pedal, gear shift handle (for forward, reverse, neutral), ignition switch, seat, windshield, rearview mirrors, and doors. There will likely be some variation for each of these, but the general location and shape and mode of action will be similar. For example, the accelerator is pressed down with the right foot – you do not force your foot under the accelerator pedal and lift up to make the car go forward faster. The color of anything in or on the car, the type of fabric on the seats, the number of doors, or whether the car is gas or electric do not matter.

So, what are the main points in both of these examples? That there are certain features, SubEntities, or Attributes that are of primary importance and these can vary to a certain extent with regard to their size, shape, or exact location, but they need to be present. So, if the Entity has a head, neck, two arms, two legs, two eyes, two ears, a nose and a mouth, whose characteristics are within a certain range, it will be recognized as another human. On the other hand, if the Entity has four wheels, a steering wheel, accelerator pedal, brake pedal, gear shift handle, etc., it will be recognized as a car. In both cases, other secondary features, SubEntities, Attributes do not matter. This is the gist of analogy.

Let's discuss a few other examples:

- **Houses** – Most of these have a certain size range and they have a roof, windows, and doors. With extremely large houses or ones that have unusual shapes, the use of analogy to identify these as a house may be stretched at times. However, once this new version of a house has been added to the "house database," the next time a similar extreme house is observed, analogy can be used to identify it as a house.
- **Dogs** – Dogs come in many different breeds. But within a given breed, no dog is identical, but each will have traits that identify that breed, such as size, shape, hair type, color, ear shape, or temperament. Each of these traits will have a range for that breed, so a given dog of that breed will usually have values of its traits that are within these ranges. For example, a Border Collie would not be expected to be the size of a Great Dane. Each dog breed has a different range for each of its traits, and these can be at extreme ends of the spectrum, for example, when comparing a Chihuahua with a Great Dane, and it may stretch the imagination somewhat to believe that these two breeds are the same species. However, all dogs have common traits, such as head and body shape, tails, foot and toe type, and personality. For example, no dog looks like a giraffe, elephant, horse, cow, moose, deer, sheep, or goat. And so, when a dog is a mixed breed, it may be difficult to use analogy to identify the dog as being a certain breed, but it should not be difficult to identify it as a dog.

- **Books** – These Entities have a front and back cover and a back spine, pages between the covers bound at the spine, and words and images on the cover and the pages. There may be a wide variety of sizes, number of pages, font type and size, and color, but analogy allows us to focus on the important Attributes (traits) and identify that a book is a book.
- **Phone Applications** – Software programmers consider analogy when they create the phone programs that we use every day. It is usually possible to download a new application on your phone, and with a little bit of fiddling around, you will be able to use it successfully in a short time without having to study an instruction manual or attend a two-day class. This is because phone applications tend to follow certain norms in how information is entered and accessed, even though these applications cover a wide range of purposes. We are able to successfully use each new application so quickly because we are using analogy, based on our experience with using other applications.

I could fill several books with additional examples, but the same analogy concepts apply to virtually every other type of Entity, including trees, roads, maps, tools, faucets, pencils, toilets, etc. Although I have been discussing analogy with respect to Entities, it also applies to Groups, Actions, Events, and MultEvents. Here are a few examples:

- **A Party** – If you see a group of people, possibly with music playing, some of the people standing or dancing, some sitting, and some talking to each other, this Event is probably a party, with is a type of Group. The number of people may be large or small, and they may be in a room, on a lawn, or by a pool, but analogy would let you determine that the Event is probably a party.
- **Baking a Cake** – This is a MultEvent that involves mixing certain types of ingredients in a bowl, transferring the mixture to a pan, and heating the pan in the oven at some temperature for some period of time. The ingredients could vary somewhat, and the size of the bowl, method of mixing (by hand or with a mixer), size and type of oven, temperature, and baking time could all vary within certain ranges, but you would still be able to use analogy to identify this MultEvent as "baking a cake."

Up to this point, I have discussed the use of analogy to identify Entities and Events that are similar to each other with regard to key Attributes. But analogy can also be used as a prediction tool. In a way, we are already doing this with the previous examples. When we see a different model car, we can "predict" that it will have a steering wheel, accelerator and brake pedals, and seat before we even see the interior.

But let's consider an example with chemistry in which analogy is useful as a prediction tool in a more complex way. Assume that we want to react propionic acid with methanol to make methyl propionate and water. We do not have any information about required reaction conditions or reaction rates and let us assume that we are not able to find any papers in the literature with this information. Let's see if we can use analogy to provide some guidance.

Propionic acid has three carbon atoms, whereas acetic acid has two. We search the literature and find that there are several papers dealing with the reaction of acetic acid with methanol to make methyl acetate and water. The papers provide information on the reaction rate at several temperatures and state that the reaction is equilibrium limited, meaning that the reaction will only proceed to a certain percent conversion, even with unlimited time.

For both of these reactions, the reactants and products have the same functional groups (organic acid, alcohol, ester, water). The only difference is that propionic acid has one additional carbon atom, compared with acetic acid. We can use analogy to predict that the reaction with propionic acid with have somewhat similar reaction rates for the same temperature with acetic acid, although the reaction rates will probably not be exactly the same. And the reaction with propionic acid will probably also be equilibrium limited, although the percent conversion at equilibrium will probably be somewhat different than for acetic acid. This will allow us to determine reaction conditions to try initially, that should provide results that are not too far from where we would like to be.

Analogy is closely related to generalization and the concept of Shared Attribute. For generalization, the key features, which are the SubEntities and Attributes, are determined for the various specific versions of a type of Entity, and the range of each of these SubEntities and Attributes can be determined. For example, if we study several different models of cars, we will discover what they have in common, and we will find these common features include four wheels, a steering wheel, accelerator pedal, brake pedal, gear shift handle (for forward, reverse, neutral), ignition switch, seat, windshield, rearview mirrors, and doors. And we will also learn that the characteristics, and location on the car, for each of these features will be similar or within a certain range. Once we have made these generalizations, we can then observe another Entity and determine if it is a car. Is this not the same as what was just discussed about analogy? So, analogy is based on generalization.

Shared Attribute for a Group of Entities refers to what Attribute, or Attributes, the members of the Group have in common. Let's consider a Group that consists of all the men in the world. Men actually have many common Attributes, but we will consider one in particular, which is that of mortality. All men are mortal. This leads to the logic example by Mill [9]:

- All men are mortal
- Socrates is a man
- Therefore, Socrates is mortal

Socrates in is the Group of all men, which we defined a moment ago, and one of the key Attributes of being in that Group is that all of its members are mortal. So, since Socrates is a member of the Group that includes all men, and all members of that Group are mortal, Socrates has to be mortal. We can apply this same logic to cars:

- All cars have steering wheels
- The Porsche Cayenne is a car
- Therefore, the Porsche Cayenne has a steering wheel

So, all cars have the shared Attribute of having a steering wheel (although, to be more precise, the steering wheel would be a SubEntity, rather than an Attribute). All members of the Group that is made up of all of the cars in the world, have steering wheels.

Shared Attribute is based on generalization, just as analogy. When we have several different versions of a given type of Entity, we use generalization to determine the key features (SubEntities or Attributes) and what the normal variation is for each of these. We can then use analogy to determine if a new Entity is a version of the type of Entity that we just generalized. And, if we know that a new Entity is a version of the type of Entity that we just generalized, then this new Entity is a member of the Group of Entities that have the key features used for the generalization, and the shared Attribute(s) are these key features.

For both generalization and analogy, we are creating rules that say the following:

- If these primary Attributes are the same or similar for a Group of Entities, set of Attributes, or series of Events, then these groups, sets, or series can be generalized. For example, "car" or "automobile" is a generalization of a large number of different models of a certain type of vehicle.
- If a new vehicle is found to have similar primary Attributes to those used for the generalization, then the new vehicle can be considered to be similar to the vehicles used to make the generalization. For example, a new model sedan from Ford or General Motors would be considered by analogy to be a car.

So, to use analogy, we first need to make a generalization of a set of similar Entities, Attributes or Events, which can then be used for the analogy comparison. And once we have defined a Group based on a generalization, we can use shared Attribute to know that members of that Group will have certain Attributes or SubEntities in common.

An AI would need to have the ability to generalize, and use analogy and shared Attribute, in a manner similar that that of a human, but it should be better at it. When comparing two or more Entities or Events to determine in what way they are similar, the AI would need to be able to compare all of the Attributes and SubEntities. To make good analogies, the AI would, ideally, have observed many of this same type of Entity or Event, therefore allowing it to know more accurately which Attributes and SubEntities are similar and to what degree, and therefore, what are the primary Attributes and SubEntities. An AI should be able to do this task of comparison in more detail, with higher accuracy, and faster, than a human for any number of examples of a given type of Entity or Event that either the human or AI has observed.

Causation

Causation, also known as "Cause and Effect," is usually associated with Events, but can also involve states, Entities, or Attributes. In simple terms, with Events, it means that if one Event occurs (cause), that a second Event will occur afterward (effect), as a consequence of the first Event. Here are some examples in which the cause is an Event:

- It rained and the grass became wet.
- The switch was flipped up and the light turned on.
- The sun shone on the dark object and the object became hot.

Often, what is believed to be causation is actually just correlation. Here are some examples:

- People becoming sunburned and ice cream melting quickly on a bright summer day.
    - These two Events are correlated.
    - But the cause of both is the sun.
- Two trees are found lying in the forest, next to each other, one morning after a storm, although they were both standing the previous day.
    - The two trees falling are correlated.
    - The cause of both to fall is likely the wind.

Often, the reason that two Events are correlated, but do not represent a cause-and-effect relationship, is because the actual cause of both of them has not been determined.

Causation also manifests itself with Entities and states, as is shown in the following examples:

- A tree casting a shadow on a building on a sunny day.
    - The tree is blocking the sun and causing a shadow to appear on the building.
        - The tree (Entity) is the cause, and the shadow is the effect.
- Water behind a dam.
    - The dam causes the water level to be high.
        - The dam (Entity) is the cause, and the high-water level is the effect.
- A bullet stopped by a bullet-proof vest
        - The bullet-proof vest (Entity) is the cause, and stopping the bullet is the effect.
- Unlocked door
    - The unlocked door allows the door to be opened by turning the knob
        - The door being unlocked (state) and turning the knob (Event) are the causes, and the door opening is the effect.

Recipes were discussed in the chapter on MultEvents and are a series of Events (with possibly some Events in parallel) that have a certain order. The goal of following a recipe is to make a product, which might be a cake. If the steps are followed precisely, or with only minor deviations, the cake will turn out correctly. So, a recipe could be considered as a type of cause-and-effect relationship, where the cause is the series of steps (adding ingredients, mixing, and baking), and the effect is the properly baked cake.

Causation allows us to predict the future just by knowing cause-and-effect relationships. If we know that it is going to rain, we can predict that the grass will get wet. We can predict that a light will come on if we move a certain switch in the right direction. And if we follow a cake recipe well enough, we can predict how good the cake will taste.

Counterfactuals represent alternative outcomes that did not occur but could have (or vice versa) under the right circumstances. We can consider counterfactuals in cause-and-effect relationships.

- The grass is dry, but if it had rained, the grass would be wet.
- The light is not on, but if the switch had been flipped up, the light would be on.
- The dark object is cool, but if the sun had shone on it, the object would be hot.

Causation requires the use of your memory to know or remember the relationship between the cause Event and the effect Event. However, counterfactuals require the use of your imagination and represent a higher level of thought. With counterfactuals, we need to be able to envision an Event that did not occur, and most likely the opposite of what actually did occur. If the counterfactual represents an Event that might not have occurred (but did), this is easier, because we are imagining a situation that did not change, and we don't have to imagine a new thing. However, if nothing happened, but we are imagining a possible Event that could have happened, a complex set of newly created images would be involved. This will be discussed in more detail in the chapter on imagination.

Just as for causation, counterfactuals allow us to predict the future, but in a different way, and can be used to avoid something bad or improve something.

- If I had applied sunscreen, I would not have been sunburned.
- If I had not left work during rush hour, my driving time would not have been so long.

- If I held the reaction condition for a little longer, the reaction conversion would have been greater.

For each of these situations, we do not know absolutely that the counterfactual that we imaged would turn out the way we expect or hope. But the counterfactual can be used to generate a hypothesis, which can be tested by carrying out the implied counterfactual activity.

As I have discussed with recipes, cause-and-effect relationships are often more complex than just one cause and one effect. Here are some of the main categories [10]:

- Common-Cause
    - A single cause has several effects:
        - A virus has one cause and several effects (fever, headache, nausea)
- Common Effect
    - Several causes converge in one effect:
        - Reduced employment, decreased currency value, increased deficit can cause an increase in government spending
- Causal Chains
    - One cause triggers an Event, which triggers another Event
        - Poor sleep leads to fatigue, which leads to poor coordination
- Causal Homeostasis
    - Causal relationship that forms stable cycle or reinforcing mechanism
        - An explosion
            - As the reaction proceeds, it creates heat and warms the mixture, which increases the reaction rate, which makes more heat…

Causal relationships form the basis for deductive, inductive, and abductive reasoning, which are related. To understand them, we must discuss the concepts of premises, conclusions, and rules. Let's consider these first with regard to cause and effect for Events. With this option, the premise would be the cause, and the conclusion would be the effect. The rule would link the cause and the effect together. Here's an example:

- Premise (cause) – It rained
- Conclusion (effect) – The grass became wet

If we had just heard, or observed, the premise and conclusion for the first time, we might not make any connection between the two Events. However, if we hear or observe these several times, we should start to create a connection and we can make the following rule: When it rains, the grass gets wet.

So, let's see what we can do with this rule. Now, if someone tells us that it rained, we can apply the new rule and conclude that the grass must be wet. We now have the following:

- Premise (cause) – It rained
- Rule – When it rains, the grass gets wet.
- Conclusion (effect) – The grass must be wet.

We can now see how these concepts apply to deduction, induction, and abduction.

- Deduction
    - The Premise and the Rule are known, but the Conclusion is unknown:
        - Premise: It rained. (known)
        - Rule: When it rains, the grass gets wet. (known)
        - Conclusion: The grass must be wet (unknown)
- Induction
    - The Premise and the Conclusion are known, but the Rule is unknown:
        - Premise: It rained. (known)
        - Conclusion: The grass is wet. (known)
            - If this happens several times, then a Rule could be created:
        - Rule: When it rains, the grass gets wet. (unknown)
- Abduction
    - The Rule and the Conclusion are known, but the Premise is unknown:

- Conclusion: The grass is wet (known)
- Rule: When it rains, the grass gets wet (known)
- Premise: It must have rained (unknown).

Based on these definitions, what can we do with each type of reasoning? *For deductive reasoning, if the premise Event has already happened, then we can use the rule to say that the conclusion Event, must also have happened.* So, even though we have not observed that the grass is wet, if we know that it rained, based on the rule that we now have, we know that if we go and look at the grass, that it will be wet. We are able know this information because the rule allowed us to determine it just by knowing the premise and the rule, but without having to actually observe the conclusion that the grass is wet. But we can also use the rule to make a prediction. If we hear a forecast that it is going to rain, we can use the rule to predict that the grass will become wet.

Induction can be used to make rules. For induction, we only know the premise and the conclusion. As was stated above, if several times, it rains and then the grass gets wet, we can make the rule: When it rains, the grass gets wet. As another example, let's say that we turned on the stove burner and then it got hot. After we do this several times, we can make another rule: When the stove burner is turned on, the burner will get hot. So, if someone tells me that they turned on the burner, I know, without actually being there, that the burner is hot.

Abductive reasoning allows us to troubleshoot or determine the cause of a problem. If we are driving our car and the engine quits, and we notice that the fuel indicator is on empty, we know that the lack of gasoline likely caused the engine to quit. This is based on the rule: If the fuel tank becomes empty, the engine will quit. Or, if the toast is burnt, we know that the cause was very likely that the setting was too high, even if we have not yet seen the setting, based on the rule: If the toaster setting is too high, the toast will burn. However, we have to be careful with abductive reasoning because there may actually be other reasons that the toast was burnt. The "bread" placed in the toaster could have been previously toasted, the bread just used for making the toast could have been less dense than usual making it burn more easily, or the toaster could have malfunctioned. There might even be rules that had already been created for each of these scenarios, such as "Less dense bread will burn unless a lower setting is used." However, abductive reasoning can at least help to generate hypotheses for the most likely causes. Often, there may be several potential causes, but one of them may appear to be much more likely than the others.

We have been discussing causation as a basis for deductive, inductive, and abductive reasoning. However, let's now consider shared Attribute as a reasoning basis. Recall the well-worn example, but rearranged slightly:

- Socrates is a man
- All men are mortal
- Therefore, Socrates is mortal

This line of reasoning was described in the analogy section as an example of shared Attribute. With this approach, we have a Group that that represents all men in the world. Men have numerous common traits, both as SubEntities and Attributes, one of which is that they are all mortal. So, if we say that Socrates is a man, then he belongs to the Group of all men, and being mortal is one of their Attributes, so Socrates must also be mortal.

However, we can also look at this example on the basis of premise, rule, and conclusion. Here the rule is: All men are mortal, and we have the following logic:

- Premise: Socrates is a man
- Rule: All men are mortal
- Conclusion: Socrates is mortal

So, if we are told that Socrates is a man, and given the rule that all men are mortal, we can use deductive reasoning to conclude that Socrates is mortal.

We could also use induction to determine this rule. If we observe Socrates for an extended period of time, he will eventually die, so we can say that Socrates was mortal. That is not enough evidence to create a rule. But, if we observe a hundred, or a thousand, or, better yet, a million men (we could do the same thing for women), they will eventually perish, and each one can be deemed to have been mortal. Now, we would be able to make a rule that states: All men are mortal, and we can use this for future deductive and abductive reasoning. So, for deductive reasoning, once someone is identified as a man, we can use this rule to state that he is mortal. But we need to be cautious here with abductive reasoning. In this case, Socrates might actually be a guinea pig, so we would be wrong to say that he is a man.

For these recent examples, we are applying deductive, inductive, and abductive reasoning to Entities, Attributes, and Groups, rather than to Events. And it may be more difficult to apply both deductive and abductive reasoning for Entities than for Events. However, this depends on how the rule is written. In the previous example, it was easy to use deductive reasoning to determine that Socrates was mortal, but difficult or impossible to use abductive reasoning to determine if Socrates was a man.

Let's consider the following example:

- Premise: Cassandra can do calculus
- Rule: Only humans can do calculus

In this case, it is easy to use deductive reasoning to say that the correct conclusion is that Cassandra is a human. However, consider the following:

- Rule: Only humans can do calculus
- Conclusion: Cassandra is a human

Here, using abductive reasoning, it is only possible to determine that Cassandra might be able to do calculus as a premise. But if the rule was: "All humans can do calculus," then we could have determined, with certainty, that Cassandra is able to do calculus.

Let's consider again the use of inductive reasoning to develop rules for cause-and-effect relationships, and we will go back to the rain and grass example. As we are collecting data about rain, there will certainly be variations in the way that the rain falls. It might be very heavy one time and light another. There might be strong winds or none at all, and there might be a lot of thunder and lightning, or none. Regarding the grass, we might study the same lawn each time it rained, or we might observe several different lawns. Even if it is the same lawn, it might have just been mowed or it might have been un-mowed for two months, and it will exhibit different levels of wetness depending on how long and hard it rained. And if different lawns were used, there are additional differences regarding the type of grass, the slope, etc.

The point here is that the data has natural variations in it, both with regard to the rain and with regard to the lawn. Analogy must be used to be able to focus on what is important. If rain is falling, it should not matter how hard or whether it is windy. If the grass has water on it afterwards, it should not matter if it is slightly wet or if there are large puddles or what type the grass is. And it should not matter how long it stayed wet before it dried. If it was wet, it was wet. No matter what kind of data is being collected, there will almost always be some variation in the data that requires the use of analogy to make sense of the data.

All rules are based on data having been collected by someone at some time. But everyone does not need to re-create the wheel by collecting their own data to develop every rule that they might use. Known rules can be obtained by reading about them or learning about them from others.

Generalization can apply to causation and the rules associated with it. Depending on the range and the way that the data was collected, the domain of a causation rule might be very specific or general. And the rule might initially be specific, but as more data is collected and more is learned, the rule becomes more general. The following chemical reaction example demonstrates this point:

- For a specific chemical reaction, it is found that increasing the temperature from 100 C to 110 C doubles the reaction rate (temperature and reaction rate would be Attributes)
    - These conditions are repeated with the same result.
    - The domain is this specific reaction at 100-110 C.
    - Rule: If the temperature for this reaction is increased from 100 C to 110 C, the reaction rate will double.
- Additional experiments (Events) are conducted in which the temperature is varied for each experiment, and it is found that when the temperature is increased by 10 C, no matter what the starting temperature, the reaction rate doubles.
    - The domain is this specific reaction at any temperature (or probably a wide temperature range with lower and upper temperature limits).
    - Rule: If the temperature for this reaction is increased by 10 C (within the temperature range used in the experiments), the reaction rate will double.
- Experiments are run with different reactions with a variety of starting temperatures, and it is found that, generally, a 10 C temperature increase doubles the reaction rate, no matter the reaction.

  - The domain is any reaction included in the experiments, at any temperature within the wide temperature range that was used for the experiments.
  - Rule: For any of the reactions included in the experiments, if the temperature increases by 10 C, the reaction rate will double.
- Finally, many different reactions are investigated, and they generally follow the 10 C reaction rate doubling, so the following generalizations can be made with a high degree of confidence:
  - The domain is generally all reactions, within the wide temperature range used in the experiments.
  - Rule: Generally, for any reaction, if the temperature increases by 10 C, the reaction rate will double.

This last example highlights the importance of understanding the domain of a given causation rule. Also, analogy would be needed for all of the experiments indicated in order to make sense of the data.

Causation relies heavily on analogy for making rules, since there would be a relatively large amount of variation in the data for most situations. An AI would need to be able to use inductive reasoning to create rules that could then be used for deductive and abductive reasoning. The creation of these rules requires that data is collected from several (or many) cause and effect examples. Each of these data sets would be somewhat different, requiring analogy to see important similarities, meaning similarities in the primary Attributes and SubEntities (and here, the Attributes of the SubEntities). Since an AI should be able to do this analogy comparison task in more detail, with higher accuracy, and faster than a human, both the analogies and the cause-and-effect relationships should be better than those created by a human.

Patterns

In terms of logic and reasoning, a pattern represents a Group of Entities, set of Attributes, or series of Events that repeats itself in such a way that interpolation can be used to determine missing elements or extrapolation can be used to predict pattern elements outside the range of available data. In some types of patterns, the pattern is a periodic repeat of the same Entities, Attribute, Event, or other pattern element based on time. For some other types, there is a predictable change, in which one could say that the pattern is based on what is the same about the difference.

Cause-and-effect usually applies to Events, whereas patterns apply to Entities, Groups, Attributes, or Events. But patterns are similar to cause-and-effect rules in two important ways. For cause-and-effect rules, several sets of similar Events are required to make the rule. Usually, each of these data sets are independent of each other, likely collected at different times, but the data sets represent a repeat of the values for the primary Attributes of the Events involved. A pattern is also based on the repetition of one or more Attributes, Events, or values, or a repetition of the predictable differences between these, although this repetition can occur within the same Entity, Group, or series of Events, or it might occur outside of it. Secondly, for both patterns and cause-and-effect, they allow for the prediction of something that is currently unknown.

Here are some examples of patterns:

1. 2, 4, 6, 8…
2. 1, 4, 9, 16… (1^2, 2^2, 3^2, 4^2…)
3. A series of Entities are placed at progressively greater distances from the previous one.
4. Entities are in a row with the following colors: Black, White, Blue, Black, White, Blue, Black….
5. Heat is applied to a mixture, causing it to react. For every 10 C increase of the mixture, the reaction rate doubles.

For the first example, each number in the series is 2 greater than the previous one, so the repeating pattern begins after starting with 2, with the addition of 2 to each series element to get the next one. This would then be the pattern rule. Notice that comparison and math was used to determine this. With this rule, we can predict that the next number would be 10. We needed enough numbers in the series to see the repeating pattern; just 2, 4…, would not have been enough. So, in this example the repetition needed to make the rule is occurring within the same series.

In the second example, we have a series of numbers, 1, 2, 3, 4…, but each of these is being squared. The repeating pattern, after starting with 1, is to take the square root of each element in the series, add 1, then square the new number to get the next element. Here we can predict that the next number would be 25.

There could be an infinite number of examples for Number 3. One non-realistic example might be that orange cones are placed on the center line of a highway, with the first two being 10 feet apart, the next one 10 feet, the next 20 feet, then 30 feet, 50 feet, 80 feet. In this case, the spacing between a pair of cones is the sum of the spacing between the previous two pairs, which is the same as the Fibonacci sequence.

For Example 4, the only math involved is that the pattern is in sets of three. We can safely predict that the next color is White. The repeating pattern is that the first color is Black, then White, then Blue, then repeat this sequence of three colors. This example shows that any type of repetition can be a pattern. There just needs to be enough repeats to be able to confidently determine the pattern. In this case one repeat is sufficient.

Example 5 was discussed in the previous section on causation, but we can also see that we have a pattern, and in fact, the statement is actually a pattern rule. If the temperature increases by 10 C, the reaction rate will double; and if it is raised another 10 C, it will double again. So, let's say that at 60 C, the reaction rate is 10 moles/second. Based on the stated rule we could create the following table:

Temperature, C	Reaction Rate, mole/s
60	10
70	20
80	40
90	80

Based on the pattern rule that we have and the table above, we can predict that at 100 C, the reaction rate will be 160 mole/second.

For each of these examples, the prediction of the next number or element is actually an extrapolation outside of the range of data that we have. I believe that we should have a high degree of confidence in the extrapolations/predictions that were made for these examples. However, we need to be careful in how far we can extrapolate. The further we extend from the data set, the higher the likelihood that the relationship or rule will break down. Often, for equations or correlations, there will be a range of conditions over which the equation applies. This could be a temperature range or be based on some other different variable(s).

In addition, we can use pattern rules to interpolate between data points or to determine missing data points within the data set. For the first example, if we have 2, 4, ?, 8, 10, we could use the rule that we derived to say, or predict, that the missing element is 6. Interpolation is, in general, safer than extrapolation, since we are working within the range of data that we have, and the pattern rule is expected to apply within that range.

There are several types of patterns, and some of the more important types are shown in the following list:

- Sequence
- Physical
- Temporal
- Series of Events (MultEvent)
- Motion
- Cyclical
- Math
- Music

Sequence

- Black, White, Blue, Black, White, Blue, Black…
- KBZ1KBZ1K…
- ♣♥♠♦♣♥♠♦♣…

Here we have colors, letters, and card symbols that are in some order that is repeating itself. We do not need to know anything about the colors or symbols or why they are in their current order, but after seeing that the same colors repeat themselves in the same order, we can make a pattern rule and use extrapolation to predict the next element in the sequence. Note that we could also use interpolation, if necessary, to fill in a missing element.

Physical

For this type of pattern, there would be a Group of physical Entities, or SubEntities of an Entity, that are arranged in 1-, 2-, or 3-dimensional order. In each case, Attributes such as distance, size, shape, or color might be used to help determine the pattern. Examples:

- 1D
    - Markings on a ruler.
    - White dashes of highway centerline.
    - Mile markers along a highway.

  - Temperature along a rod that is heated at one end (at steady state). For this case, there would be a change in temperature, from one point to the next, that would be the same if the correct mathematical function was applied.
- 2D
  - Decorative picket fence
  - Floor with regular pattern
  - Wires on a screen
  - Chess board (just the squares)
  - Wallpaper
- 3D
  - Stone blocks for the Egyptian pyramids (at least on the outer surface).
  - Frame for a geodesic dome.
  - Floor plan for each of the stories in an apartment building.

Temporal

Several Entities of the same kind could change with time in a similar way. A pattern rule could then be made that would allow the prediction that other Entities of the same type will change in the same way. Examples:

- Humans aging.
- Automobiles aging.
- Knowledge increase of students as they progress through Grades 1-12.

None of the aging humans, cars, or students will change in the exactly the same way, so analogy and generalization need to be employed with the collection of the data in order to create the pattern rules.

Series of Events (MultEvent)

- Following the same recipe every Thursday.
- Mass production of a product.
- Running the same chemical reaction experiment more than once.

In each of these examples, there is likely not a pattern imbedded with the process. However, if each of these procedures is followed two or more times with the same result, a pattern emerges as follows: Follow procedure – Obtain desired product.

Motion

- Canada Geese flying in formation.
- Trajectory of a certain type of projectile.
- Time for a given aircraft to reach take-off speed.

The geese example is more complex than it seems at first. The V-shaped flying formation has patterns related to the distance between each goose and the overall angle of the "V". Also, the geese will fly at a speed that is within a range for Canada Geese. This pattern will apply to most other flocks of Canada Geese. So, the patterns are defined within the Group. However, by observing several flocks of geese flying in approximately the same formation, we can conclude that all geese fly in this way. We can now make two conclusions about the flying patterns of geese. (1) We know the main details of the V-formation, regarding bird spacing, angle of "V", and the average speed of the V-formation and (2) We know that it is not just one flock of Canada Geese that flies in this manner, but all flocks of this species.

Analogy, generalization, and probability are all involved with the ideas just presented. Each goose is slightly different than all others, but the same with regard to the primary Attributes. The distance between each goose in formation will be slightly different, and even vary from one second to the next, but this distance will be within a certain range. The angle of the "V" will vary slightly and the number of birds in each flock will vary, but these will both be within a certain range.

So, we are using generalization to define the mean and range of each of these variables, and the application of analogy allows us to say that a given flock and flight pattern is similar to other Canada Geese flocks and flight patterns. In addition, after observing many flocks, we have concluded that all Canada Geese fly in approximately the same V-shaped pattern. However, there is some probability that we might find a flock that flies in a pattern that is outside the norms of most Canada Geese.

Regarding the second two examples above, these could also be included in the math section.

Cyclical

- Moving parts of an engine.
- Rotation of an Entity in the clockwise direction of 90 degrees every minute.
- Motion of a swinging pendulum.
- Motion of the planets in the solar system.
- Day-night cycle.
- Seasons.

Math

There are many kinds of math, so also many kinds of patterns. Just a few examples will be listed here. In some of the examples listed above, math can be used to model what is occurring, so some of these examples will be listed again here.

- Cone
    - Starting at the apex, for each incremental movement downward, the diameter increases by the same amount.
- Geometric drawing with repeating images.
- Temperature along a rod that is heated at one end.
- Trajectory of a certain type of projectile.
- Motion of the planets in the solar system.

For each of the last three math examples, an equation could be derived that would describe the temperature profile, trajectory, and planetary motion. These equations might be derived by observing data from experiments, or by using engineering and physics, and they represent a pattern for each example. In each case, the equation could be used to interpolate and determine the conditions at different times or locations. The equations could also be used to extrapolate and predict the result for a longer rod, the case of the projectile being propelled from a higher location, or the location of a planet several cycles in the future.

Music

After listening to a song one or more times, we will usually recognize the melody and some of the words if it plays again. If we do recognize the song, we should be able to predict how the song will progress. Also, the chorus of modern songs will usually repeat itself at least once, so we are able to hear a pattern for part of the song before it even finishes.

There are two important concepts being presented here. First, some of the patterns are based on a set of letters, numbers, symbols, shapes, colors, etc. that have a certain order that repeats. This compares with patterns that are based on a change from one pattern element to the next in which the change is the same in some way. For this second type, some form of math is involved to represent the change, which may be simple arithmetic or a complex function, but in some way, the change is constant between the elements. So, for some examples, the pattern is found within the Entity or Event, while in others, the pattern is determined by observing several or many of the same types of Entities or Events in which their order repeats.

Now let's compare physical and temporal patterns. For physical patterns, there is either a repetition within the Entity or Group, or a mathematically predictable change, that allows the pattern to be determined. But, for temporal patterns, the pattern is determined by observing the changes that occur with several or many examples at different times of the same Entity.

Hierarchy

Often patterns are hierarchical. One example of this is the day-night cycle. Each day there will be a certain number of hours of day and night that approximately repeats itself every 24 hours. However, as the year progresses, the number of hours for day and for night changes, and this cycle takes 365 days. But we also have the temperature change with day and night that varies with the season. So, at a given point in the seasons, and at a specific location, there will be an average high and low temperature that can be plotted on a graph. However, as the year progresses, the plot will trend up or down, and there will be a year-long cycle overlayed on the daily cycle. In addition, the temperature variation at both levels is caused by the day-night cycle, meaning that we also have causation involved.

We can look at this same example from the perspective of the Earth's rotation and revolution around the sun. Now, let's include the moon, also. The Earth orbits the sun in roughly 365 days and rotates once for each of those days. In addition, the Earth's axis is tilted 23.5 degrees from the perpendicular, relative to its solar orbit. Let's consider a point in the northern part of the United States, for example, somewhere in northern Idaho. As the Earth rotates, on a given day at noon, the sun will be observed to be at a certain angle from the horizon, and the length of the day (as daylight hours) will be a certain number of hours. But due to the 23.5-degree inclination, each day, the sun's angle from the horizon at noon and the number of daylight hours changes by a small amount. Therefore, there is a cyclical pattern for each of these that is caused by the combination of the periods of rotation and revolution, and the 23.5-degree inclination. For each point on the Earth's surface, the cycles would be slightly or a lot different. Since we understand these patterns, we can predict the amount of daylight and the sun's angle from the horizon at any point on the Earth 100 or a million days in the future.

Now, let's add the moon with its orbit to this discussion. The moon revolves around the Earth in 27.3 days. If we were to look down on the orbits of the Earth and the moon, we could draw an imaginary line representing the complete path of the Earth in its orbit. The moon would cross back and forth across the Earth's orbital path as the moon revolved around the Earth. If a line could be drawn of the moon's location, the path would resemble a curved sine wave overlayed on the Earth's orbital path. This interesting pattern is caused by the hierarchical relationship of the orbits of the Earth and the moon.

If we look at a chessboard (without the pieces), we can consider what the pattern rules might be that describe the layout. All standard chessboards have 64 squares, with half of them a dark color and half of them a lighter color. For our board, we will use black and white. Here are the rules that describe the pattern for the layout of the squares:

- Entity
    - 2 Dimensions
    - 1 Flat square
    - SubEntities
        - 2 Dimensions
        - 64 Flat squares
            - Entity divided into 8 equal rows and 8 equal columns.
- Pattern Rules
    - Only 2 colors, in this case Black and White.
    - Lower left SubEntity is Black.
    - For each SubEntity, all four bordering SubEntities are the opposite color, but these four are the same color as each other.

These rules should allow any chess board to be colored with the correct pattern. The overall size is not one of the criteria, although most chessboards are within a certain size range. Any two colors can be used, although the colors are usually dramatically different from each other to provide contrast.

Now let's look at an example from Brain Metrics Initiative [11] that is typical of a pattern recognition problem from an IQ test (see Figure 7).

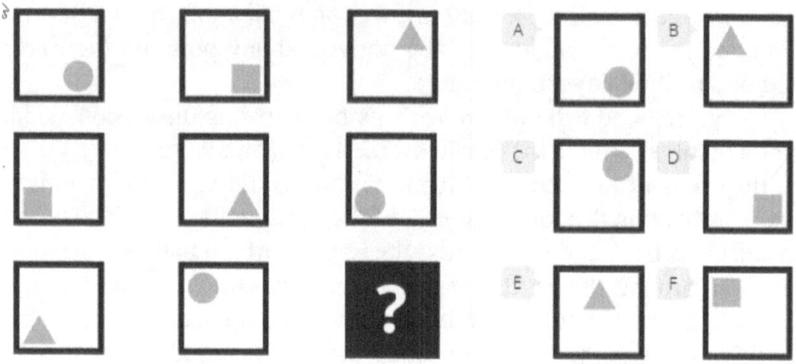

Figure 7. IQ test question example.

On the left side we have nine squares, eight of which have a circle, a square, or a triangle located in one of the four corners of each of these eight squares. We need to determine which is the correct shape in the correct corner for the square with the question mark using pattern recognition. On the right side are shown six possible answers.

To solve this problem, we first need to determine how to group the squares to look for patterns. We would rule out grouping all eight of the known squares together because this does not allow for the repetition that is needed to determine a pattern. At the other extreme, having eight groups of one does not work either because the changes from one square to the next appear to be random.

The groupings that appear to make the most sense are groups of three divided by rows or by columns. This would allow us to see a potential pattern, with one complete repetition and one partial repetition. We are assuming that a pattern can be found in a group of three.

There does not seem to be an obvious reason to choose rows or columns, so let's start with groups based on rows, and we will look at each row from left to right since we are accustomed to doing so when reading and writing. In the top row we see a circle in the first square, a square in the second, and a triangle in the third. Each of these is located in one of the four corners of the square, but we do not see any patterns at this point that would allow us to determine the reason for the specific placements.

For the middle row, we have a square, triangle, and circle in Squares 1, 2, and 3. The square and the triangle are in the same order, but the circle is in the last square, rather than the first. So, a possible pattern is that whatever symbols were the second and third in the previous row become the first and second in the current row and the first symbol moves to the end position. Regarding the symbol locations, we observe that each symbol has moved clockwise to the next corner in it respective square, when comparing the top and middle rows.

We appear to have two independent patterns that represent a sort of hierarchy. The primary pattern is the order of the symbols, and the secondary pattern is the location of each symbol within their respective squares. When we look at the bottom row, we see that the first two squares adhere to both of these patterns. The unknown square must contain a square symbol, and since the square symbol in the middle row was in the lower left corner, it must be in the upper left corner for the unknown square (based on the clockwise movement). So, the correct answer must be "F".

Now let's change the groupings so that we compare columns, starting with the left column, and moving to the right. We see the same pattern for the order of the symbols, so we can conclude that the unknown square has a square symbol. However, we do not see consistent changes in the positions of the symbols. The triangle and the square consistently move counterclockwise, but the circle first moves diagonally, then counterclockwise. So, we do not have a consistent pattern regarding symbol placement, and, therefore, it is not possible to determine the placement of the square symbol in the unknown square. We have determined that the only way to solve this problem is to group the squares in rows, as we did originally.

This problem had two patterns that were both very ideal in that there was no randomness involved. Real problems will usually have some randomness that will make it more difficult to see the pattern and will require the use of generalization to assist in recognizing the pattern. And there could be higher numbers of patterns involved, increasing the complexity. As the complexity increases, an AI should be able to be more capable than a human at discerning patterns, because the AI could have more memory and processing speed, so the AI should be able to keep track of more details and evaluate the various possible patterns faster and more accurately.

Mathematics

Mathematics is essentially a form of logic because it is able to determine conclusions (calculation results) based on rules (mathematical operations) applied to inputs (numbers). As for other forms of logic, it allows us to make known something that was unknown using things that are already known. With regard to the concepts presented in this book, and application to an AI, I think that the main value of mathematics is its use in helping to describe and understand the world, but especially the physical world.

There are many areas of mathematics, but some of the most important are arithmetic, algebra, geometry, calculus, statistics, and numerical analysis. Each of these can be sub-divided into additional disciplines, and there are, of course, other areas not mentioned here that are of high importance. Each of these areas has its own rules and levels of complexity.

Math has its own value as a form of logic. However, it also plays an integral role in the other forms of logic, to help express correlations and causation, patterns, comparisons, analogies, and generalization.

Probably our first exposure to math is with arithmetic. We learned and memorized addition and multiplication tables, based on the numbers 0 to 9, which were essentially a set of rules. With additional rules we learned how to add and multiply numbers of any size. For example, if I have 5 apples and 7 oranges, and I apply the addition operator (rule), the result is 12 pieces of fruit. We could just count all of them to get 12. That's a little more difficult, but not too bad. However, it does allow us to prove that 5 + 7 is equal to 12. But let's say that we have 135 apples and 589 oranges. Now, we see that it is much easier to apply the addition operator to these two numbers than to count all of them.

An AI solving this problem would not just need to be able to do this calculation. It would have to understand that 5 represents these many Entities, Groups, or Events (or Attributes) and the same thing with the number 7. And after the addition operator had been applied to these two numbers, the resulting number 12 also represents this many Entities, Groups, or Events.

For multiplication, the expression 5 x 7 would mean that we have 5 Groups, and each Group has 7 Entities. An example might be 5 rows of squares with 7 squares in each row. As above, we could count all of the squares to determine the total to be 35. Or we would use the multiplication operator on 5 and 7 to get 35.

Now let's look at a right circular cone, a 3D geometric shape. To find the volume of the cone, we could submerge it in a tank of water just large enough to fit the cone, then measure the volume of water displaced by the cone using a calibrated vessel. However, this is a tedious task that is prone to significant errors at several points. If we know the radius of the base and the height of the cone, we could just use the equation for the volume of this type of cone:

$$V = \frac{1}{3}\pi r^2 h$$

Here, we are applying multiplication, division, and exponential operators to accomplish this task in a manner that is much easier and more accurate.

Now, let's consider some of the ways that math is utilized to help understand the physical world. Assume that we are collecting data on plant growth for a certain species of plant. We have 9 test subjects, of which each of them has just sprouted. We measure the height every 5 days for 60 days and plot the results, which are shown in Figure 8:

We can see that there is a fair amount of scatter due to the different heights for each of the plants, as would be expected, even with identical conditions for each plant. The light gray points represent the average height for each data set (at a given time). A 3rd order polynomial equation (example of algebra) was fit to the average heights and the equation is:

$$H = 0.0002T^3 - 0.0494T^2 + 4.0056T - 3.2283$$

$$H = Height\ of\ Plants, \text{cm}$$
$$T = Time, days$$

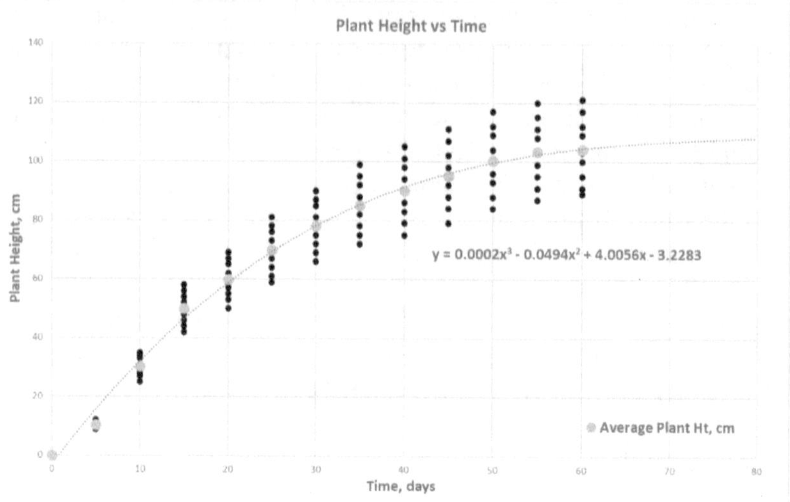

Figure 8. Plant height vs time.

By determining the average value at each time, we are essentially generalizing the plant height for that time, meaning that there will be a certain range of heights for this plant species, under the conditions used for the experiment, but that we can expect the values to be around a certain average value (we are using statistics here). This is the same concept as when we generalize other Entities, such as a human face. With the plant, we know that it will grow, and for a given time period, its height will be in a certain range. With a human face, we know it will have a nose, and the size of the nose will be within a certain range.

As discussed above, we can fit a line through the average height values. It may not be a perfect fit, but if it is reasonably close, such as we have in the figure, we can do several things with it. The polynomial line represents a pattern. And this pattern allows us to interpolate between the data points and extrapolate past 60 days. So, we can use the equation to estimate what the average plant height would have been at 25 days. We can also predict what the height should be at some time after 60 days. But we need to be careful about how far in the future we can accurately predict. Probably 10 more days, maybe 20. But I would not trust the prediction past some maximum number of days (maybe 80 total days is a good estimate). The ability of fitted equations to accurately predict values outside the range of data depends upon how good the fit is and how linear the equation is. Very generally, equations that are linear, or nearly so, will be able to accurately extrapolate further outside of the dataset than equations with some curvature.

With this equation, if the experiment was replicated, we should be able to accurately predict what the average plant height will be for any day up to 60 days, and for several days beyond 60.

Now let's look at how math can help us with a different kind of variation in the data. In the chemical industry, distillation is commonly used to separate compounds based on their boiling point difference. If you boil a mixture of two compounds with different boiling points, the concentrations of each of the compounds will be different in the liquid and vapor phases. This difference is important for determining how difficult it will be to separate the two compounds from each other using distillation (how large does the distillation column have to be and how much energy will be required)

One type of data that is useful for this purpose is the following. A mixture is created with a specific composition, the temperature is set at a certain value, and then the pressure is adjusted so that the mixture is at its boiling point (so there is a vapor phase in equilibrium with the liquid phase). The pressure is determined and both the liquid and vapor phases are analyzed to determine the composition. The liquid composition is then changed and the procedure repeated.

Figure 9 shows plots of data collected, as described above, for ethanol and water by Matugi [12]

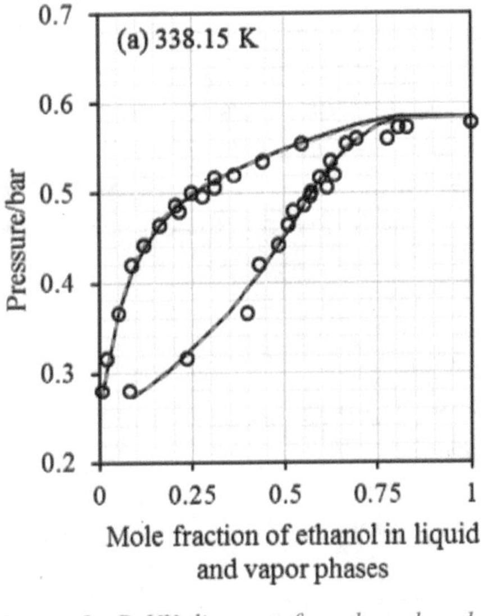

Figure 9. P-XY diagram for ethanol and water.

Equations have been fitted to this data and are overlayed on the data points. As can be seen, there is some scatter in the data, so most of the points are on one side or the other of the lines representing the equations.

For the previous example, there was variation in the data at each time because the plants used in the experiment naturally grow at different rates. So, no matter how accurate the height measurements were, there would still be variation in the data.

For the current example, we have a different problem. It is difficult to obtain extremely accurate analyses of vapor and liquid phases due to the nature of the analytical instruments used for this purpose. If we had perfect analyses, and enough data points, the data points would create very smooth curves. However, due to the inherent inaccuracy of the analyses, the equations that have been fit to the data are likely providing more accurate compositions, for given temperatures and pressures, than the actual analyses. And the equations for the liquid and vapor phases are also enabling the use of interpolation to determine compositions between the data points.

So, here we are using math to improve the accuracy of our data and to determine vapor concentrations at temperatures, pressures, and liquid concentrations where data has not been collected. In addition, if this experiment were to be repeated, we could then predict the results for any set of conditions that had been used in the first experiment.

For the final example, obtained from Riddle [13], we will use calculus to allow us to predict the growth of bacteria. It is well known that the growth rate of bacteria is directly proportional to the size of the colony. For this example, we have 5000 bacteria at noon and 10,000 at 2 PM. We want to know how many we will have at 5 PM.

We will start with the following equation based on calculus:

$$\frac{dx}{dt} = kx$$

$x = number\ of\ bacteria$
$t = time, hr$

This equation can be integrated to obtain the following equation:

$$x = 5000e^{0.3466t}$$

With this equation, we can predict the number of bacteria that we will have at any future time. For example, at 5 PM, we will have 28,300.

By utilizing calculus, we are able to use only two data points to derive an equation that can estimate how many bacteria were present before 2 PM (interpolation) and can predict how many there will be at any time after 2 PM (extrapolation). Of course, just as for the previous examples, we need to be cautious about how far into the future we try to predict. The accuracy would decline as the time into the future is increased.

The purpose here has not been to demonstrate how to solve math problems, but rather to show how math can be used as a form of logic and reasoning to fill in information that does not exist, correct or generalize existing information, and help predict what might happen in the future.

There are now many tools, in the form of calculators, spreadsheets, and software programs to help solve math problems. Students are normally taught how to solve math problems by hand, so that they understand each type of math, but eventually come to rely on calculators, spreadsheets, and software programs as they progress through their education and eventually become employed as engineers, scientists, accountants, teachers, medical professionals, etc.

I believe that an AI should do the same as a human in this regard. An AI should be able to use math to set up and solve real physical problems. For example, it should be able to understand a physical situation, determine what type of math is required, set up the algebraic, trigonometric, differential, etc. equation, and then solve it. For differential equations are too difficult for an analytic solution, it should know how to use numeric methods.

By saying that an AI should know these things, it could either learn them in the same way as a human, or this ability could be programmed into the AI as an innate ability. I will discuss the innate abilities that an AI should have in the next chapter. However, for solving most math problems, one option would be for its equation-solving abilities to be built into its software. However, it might make more sense to just allow the AI to have access to the same kinds of software as a human does, such as spreadsheets, programs such as Mathematica® or MATLAB®, process simulators, etc., and have the AI learn how to use them in the same way that a human does. The AI could have direct access to these programs and, therefore, be able to interface with them more efficiently than a human can.

One option would be for the AI to just rely on these existing software programs to solve its math problems, without having an understanding of how to do the math itself. However, I believe that this would limit the AI's abilities to truly understand a given physical-based problem, apply the correct math technique, and then understand the results and how to make use of them.

Imagination

Imagination is the human mind's Play-doh®, Lego®, and Erector® set. It allows a person to transform any Entity, Group, Event, MultEvent, or Rule stored in memory in any way that it wishes. It is a critical part of learning and prediction. But it is highly dependent on a person's experiences and existing knowledge.

Using imagination, the Attributes of an Entity can be mentally changed, the relationship between Entities in a Group can be modified, Events can be revised in any way desired, and the order of Events in MultEvents can be rearranged. For example:

- Entity Attributes
    - Size
    - Location
    - Orientation
    - Velocity
    - Age
    - Any other Attributes
- Group Entity Associations
    - Order
    - Relative Distance between Entities
    - Orientation of one Entity relative to another
    - Relative size between Entities
    - Relative velocities
- MultEvent Events
    - Order of Events
    - Addition or removal of an Event
    - Timing of one Event relative to another
    - Speed of one Event relative to another

The rules for logic can also be modified as a type of thought experiment. These imagined rule changes may or may not be feasible, but can assist with thinking creatively about a task, in the same way as for imagined Entity, Group, and Event changes.

Imagination relies on the information that is stored in memory as a basis for the modifications. Essentially, a base case is needed to have something to make a change to. This is why it is important to see and experience as much as possible and to read to learn about things that cannot be personally observed. It is also important to spend time doing and creating things, since these activities provide analogies for how to make modifications to other things.

Our imagination starts with the retrieval of some Entity, Group, Event, MultEvent, or Rule from memory, and our use of imagination may be no more involved than that. Simply "pulling up" a memory. But we may also make some changes to this memory. In this case, we not only need a base case (the original memory) to start with, but knowledge of the types of changes that are possible. For example, we might have a bolt that we want to screw into a set of threads in a hole of a metal plate. We will likely need to rotate the bolt in order to line it up properly for it to line up to be threaded in. We are starting with a bolt that we have seen and are aware of. In our minds, we can rotate the bolt until it aligns properly. But even the mental activity of rotating the bolt is based on having observed the rotation of other Entities. We can create new Entities, Group arrangements, or Events and MultEvents that have never existed, but both the starting Entity and the type of modification must already be known.

Let's look at some examples of imagination manipulation:

- Entity
    - Part for an engine is rotated in one or more of 3 axes.
    - Chess piece is moved to another spot on the board.
    - Wood block is enlarged.
    - Color of a house is changed to another color.
    - Old human is modified to be a young human.
- Group
    - Engine part is moved closer to another Engine part.
    - Small box is placed inside of a larger box.
    - Engine part is rotated relative to another engine part.
    - Number of Entities in a Group is increased or decreased.
    - The words in a sentence are rearranged.
- Event
    - A recipe is changed so that twice as much can be made.

    - The order of two SubEvents is reversed.
    - A new step is added to the middle of a procedure.
    - A song is performed faster than the last time.
- Rule
    - A reaction rule that says that a reaction rate doubles every 10 C could, instead, double every 5 degrees.
    - A rules states that stepping on an Auto accelerator will make the car go faster, but it could, instead, make the engine quit.

Imagination also works hand-in-hand with the various forms of logic. If someone uses analogy to explain how a new Entity is similar to something that you are familiar with, your imagination will immediately recall the familiar Entity and mentally make the modifications to it that will then allow you to understand the new Entity. If you see a half-finished jigsaw puzzle, or a written sentence with a non-legible word, or part of an animal that is behind a tree, your mind will use imagination and memory and/or pattern recognition to fill in the missing parts.

When reading a novel, each of the words is actually shorthand for an Entity, Group, or Event that you may or may not be familiar with. You may read about a description of mountains that are shaped in a unique way with a type of tree that you are not aware of. You have probably seen mountains and trees, but they are different than those described in the book. As you read, your imagination will retrieve mountains and trees that you are familiar with and then proceed to modify them based on the descriptions on the page. As you continue through the book, your imagination will create an entire world and store it in your memory. Reading adds to your knowledge, but just as significant, it exercises your imagination.

So, imagination is an integral part of most of the thoughts we have from the time that we wake up until we go to sleep. But I would like to categorize somewhat how we most commonly use imagination:

- Understanding past Events.
- Considering counterfactuals.
- Predicting what will happen in the future.
- Evaluating scenarios and predicting what will happen for each one.
- Determining what changes should be made so that a desired outcome occurs.
- Assembling Entities.

- Creating a new Entity

For an AI to achieve AGI, it will need to have these imagination capabilities. It is the key to understanding the physical world and being able to be creative within it and make predictions about what will happen based on a given set of conditions. But an AI should be able to have a much more capable imagination than a human. Compared with a human, an AI should be able to remember everything in greater detail, make more complex mental revisions, integrate these mental revisions with other associated revisions, and then remember the full details of all of these activities.

6

Innate Concepts

"All progress is based upon the universal innate desire on the part of every organism to live beyond its income."

- Samuel Butler

For humans, there are knowledge and abilities that are either innate or are learned at an early age. A newborn baby, for example, knows how to eat, drink, open its eyes, smile, laugh, and cry. These abilities are truly innate. But I believe that there are other concepts that are innate but cannot be expressed until a young human has learned some practical aspects of the world, and there are other concepts that are not innate, but are learned at an early age. The following list includes some of the most important basic concepts:

- 3D space
- Location
- Time
- Motion
- Force
- Cycles
- Sequence
- Hierarchy
- Probability
- Possessions
- Quality
- Value
- Effort
- State of Being
- Existence
- Emotions

We very early understand 3D space and, after bumping our heads several times, we learn that two physical objects cannot occupy the same space at the same time. We learn the locations of things we do and do not like. When we have to wait for something we like or we learn how long it takes us to do some task, we learn about time. Motion is all around us; people walking, dogs running, birds and insects flying, cars moving. Our first experiences with force are probably gravity (when we fall or drop something), and then when we push or pull on some object.

We learn about cycles by our daily routines (get up in the morning and go to bed at night, as well as breakfast, lunch, dinner) and the day-night cycle and the seasons. Our early experiences with sequences include the procedures for getting dressed in the morning and getting ready for bed at night, as well as many activities during the day. Probably our first experience with hierarchy is with parents and children, and maybe between humans and pets.

Probability is all around us from an early age. Things drop, they break or do not work, things do not happen when you expect. We learn that some things are our possessions, such as our clothes and toys, but other things (the toy that we wish was ours) belong to someone else. We learn about the concept of quality with our toys that used to work great when they were shiny and new but now are tattered and broken. We find that some food is delicious and so we think it has higher value than the food we hate, and some toys are more important to us than others (and therefore have higher value).

We learn that it takes more effort to climb up stairs than to walk across a level floor, and big things are usually harder to lift than little things. Solid Entities are all around us from birth, we drink (and spill) milk and splash around in water, the wind blows leaves, paper and our hair, and we see bugs and dogs, grass and trees, and stones and dirt, all of these teaching us about state of being. When we have a toy for a while then lose it or it is thrown out because it no longer works, we are learning about existence (it existed and now is non-existent).

When a baby cries, smiles, or laughs, it is obvious that the related emotions are innate. At an early age, we develop more sophisticated emotions, such as love, hate, joy, fear, generosity, jealousy, etc.

We even learn about some chemical engineering concepts at an early age, although many of us will not recognize them as such even as adults. When we feel hot or cold water, we are experiencing heat transfer. The covers on our bed make us feel warm because they insulate and reduce heat transfer. The warmth of the sun, or a burning fireplace, is due to radiant heat transfer. Peanut butter or a thick milkshake demonstrate liquid with high viscosity. Water forms large droplets, and Water Strider insects and needles can float on water, due to their high surface tension. And when we get burned by the uninsulated handle of a hot pan on the stove, we are experiencing the fact that metal has high thermal conductivity.

As humans grow and experience the world, they quickly add to their understanding of how the world works and build on their innate abilities and concepts. However, an AI has no innate concepts except those that are programmed into it. And we have to consider the various potential levels of interaction that an AI might have with the physical world. An AI might be a program running in a computer (or many interconnected computers), in which it has no interaction with the physical world. The other extreme might be an intelligent robot with multitudes of sensors to replicate the five senses of a human, and possibly more (ultraviolet/infrared, microwaves, radio waves, ultrasound), that would allow it to experience its surroundings, conduct experiments, and learn causation relationships. But it, too, would only have the innate concepts that are programmed into it.

For an AI to have AGI, it would certainly need to be able to reason and use logic, as discussed earlier. However, it would also need to be able to learn about new Entities, Events, and logical relationships. Some number of innate concepts would need to be programmed into an AI at its beginning. However, this could be the bare minimum amount, so that it starts its "life" like a human baby. It could then learn physical relationships through interaction with the world and be taught information through written, and possibly verbal, communication. Its innate knowledge and understanding could allow the use of analogy to learn this information. For example, if the AI understood the characteristics of a young adult, it should be able to use analogy to learn about a child or an old person. Or if it knows what a dog is, then it should be able to understand a horse.

An AI should be able to learn written information at a much higher rate than a human, and acquisition of the verbal information should also be faster since less repetitions would likely be required. If the AI also had sensors and actuators of various types, it would be able to learn some physical aspects of its surroundings, as well as causal relationships.

At the other end of the spectrum would be to program in as much information and capabilities as possible. Every known word and the grammar rules of a given language and a complete understanding of almost all branches of mathematics could be programmed into the AI. All knowledge of biology, chemistry, physics, every branch of engineering, all of history, could be embedded. Then the AI could go on to learn new information that becomes available, as well as learn new causal relationships and patterns, using its phenomenal educational background as a starting point.

This last option sounds like the most desirable one, on the surface. But if the AI is a self-contained unit, how large would the computer need to be and how much energy would it consume? Another option would be to have a central computer with all of this information, that numerous AI devices access by remote communication. In this case, the AI might have knowledge and mental capabilities similar to a human and be able to access the central system of computers, which carries out the mental calculations. This would be similar to a human accessing the internet.

The answer will depend on the purpose of the AI and the programming languages used. If the purpose of the AI is to act as an engineer, then it probably does not need to know a lot about world history. However, we may want to have a system that is somewhat like one of the current large language models (LLM), in which we can ask it anything.

At this time, I would like to focus on the cognitive aspects of AI without robotics or the associated physical actuators. Based on this point, it seems that there are at least three main scenarios which would be the most useful. One would be the LLM-style system, that is one massive central computer with knowledge of everything, that is remotely accessed by millions of AI devices. Another version of this concept would be a network of many specialized systems, and when a request comes in, it is directed to the appropriate specialized system. In both cases, the local AI system would have limited capabilities. A third version would be one in which the individual AIs have capabilities similar to humans, or maybe somewhat more advanced, but are able to access a centralized system similar to the internet. For any communication interruptions, this third option would allow the AIs to continue to function much in the same way that a human can even when it loses internet access.

The third scenario would be one in which the local AI device is like a very well-educated human that knows everything about one or two fields of knowledge and general knowledge about most other things. And it would be able to access the internet or a central computer system to search for information when necessary. For example, the purpose of the AI might be to act as a chemical engineer for a US company. It should have complete knowledge of the English language, advanced abilities with math, and all available knowledge of chemical engineering. And it should have general knowledge of many other areas, such as politics, history, geography, other engineering disciplines, the environment, how to interact with humans, etc., and the AI would have advanced reasoning capabilities.

With any of these scenarios, assuming that the AI was not an embodied humanoid robot, I would envision that the AI would interact with people as if it were working in a remote location, and communicated through video call systems such as Zoom, Teams, WebEx, etc. Therefore, it would have access to the camera and microphone of the computer that belongs to the person interacting with the AI. And the AI would appear on-screen as if it were an actual person. In this situation the AI would function as a team member working remotely.

Now, as an example, I would like to discuss the idea of an AI being employed as a chemical engineer, and the knowledge and capabilities that the AI would need to have (these capabilities would apply to most technical fields that deal with the physical world). To begin with, it should understand the basic concepts presented at the start of this chapter. Regarding its 3D capabilities, the AI should be able to mentally manipulate Entities and Groups in the same way as a human. It should be able to retrieve Entities from its memory and move them at different velocities along trajectories and relative to each other, and rotate, enlarge, resize, reshape, combine, or dismantle them, then save the revised Entities to new locations in memory. It should be able to retrieve Events and replay, modify, rearrange, combine, add or subtract other Events, then save the modified versions.

The AI should be able to use the camera, or videos or photos from the internet or through communication to obtain new information about Entities and Events in the same way that a human would. It also should be able to read text or listen to verbal communication and generate mental images of the Entities and Events that are portrayed, and store these in memory, so that they could later be retrieved and used as described above.

Another basic concept that the AI should understand is that of absolute or relative location in 1, 2, or 3 dimensions. An AI should be able to understand and visualize locations and Entities being in locations. It should be able to study a globe or map and understand how to use it to go somewhere or pinpoint its current location on a map.

Like location, time can be absolute or relative. As humans, time is deeply ingrained in how we think. What time do I need to get up, what time do I have to be at work, how long will the commute be this morning, how long is that meeting I don't want to attend, how long will it take me to do that project. The AI should have a similar understanding about Events and when they have or will occur and know the length of time for an Event. Based on its knowledge database, as well as new time-related information, the AI should be able to use analogy to estimate how much time a new activity might require.

The AI should also be able to understand motion (how fast is that car going, the motion of a flying baseball, a bird in flight), force (effect of gravity while getting out of bed, loosening a rusty bolt, pushing a loaded shopping cart), cycles (getting up and going to bed, the four seasons, the rotation of an engine), sequence (the alphabet, recipes, operating procedures), and hierarchy (family tree, company organizational chart, book structure of chapters, paragraphs, sentences, words).

Probability is ingrained in language with words like if, could, should, would, and subjective verb tenses, but also with the concepts of contingency plans, battery backups, spare parts, understudies for plays. Possessions are very important to us, and we concern ourselves with our salaries, net worth, the size and location of our house, whether our car is a luxury model or not, our wardrobes, what "toys" we have (motorcycles, boats, snowmobiles).

We understand quality in two ways: how well-built and functional something is, and whether some device works well or is broken (or brand-new vs junk). The value of something can either represent the amount of resources (money) required to obtain it or its quality relative to what it cost.

Effort can be tied to force, time, or both, and is expressed by a project that takes a long time at work, moving the refrigerator, fixing the car, planning a trip with a complex itinerary. We experience state of being with ice, water, steam, thousands of physical objects we interact with daily, flowing water in rivers and waterfalls, and breezes and windstorms, but also the animals, plants, and inanimate objects (rocks, buildings, cars). We understand the concept of existence as the current officemate, the neighbor that died, the imaginary character in a novel, and the bridge that was never built.

An AI employed as a chemical engineer should understand and experience most of these concepts in the same way that a human does, but it should also be able to understand how a human would think about them. The AI may not have a physical body or human-like emotions, but it should be able to relate to them. For example, from a physical point of view, the AI should understand how a human can move, generally what it can do, and how much weight it can lift, even if the AI itself is not capable of doing these things. If the AI is designing a process, suggesting experiments to run, or requesting modifications to existing equipment, the AI needs to understand what a human can do in order to not suggest activities that are impractical or infeasible. If the AI is truly going to act as an AGI, it must comprehend how humans think, what they can do, what is important to them, and their emotions, on a general basis as well as for specific individuals.

Regarding linguistics, the AI should have a complete knowledge of English if working in the US, and possibly other languages. If employed in another country, then a language other than English should be its primary language. The AI should have a complete knowledge of the primary language, including most of the words and their meanings, as well as grammar and idioms. However, the AI should be able to learn new words, no matter the part of speech, which would include technical names for Entities, company names, abbreviations, and, also, idioms. The AI should be able to create the 3D images (for Entities) and 3D videos (for Events) for each of these.

The AI should understand mathematics at a reasonably high level but would not need to be an expert in every branch of math, just as most chemical engineers are not. Being able to solve problems using arithmetic, algebra, geometry, trigonometry, calculus, differential equations, matrices, numerical analyses, and a few other branches of math would be sufficient in most cases. But, just as for humans, the AI could and should make use of the same software tools that a chemical engineer would. For example, spreadsheets and chemical process simulators. So, the AI should understand the physical nature of a problem, what type of math is required to solve it, and for simple problems, perform the calculation itself. For more complex problems, it should know what software to use, how to use it, and it should understand what the results mean. For any of these problems, the AI should understand how the answers relate to the real physical world. In addition, the AI should be able to learn about new software and learn how to use it. For problems beyond its capabilities or those of a central computer system, it should be able to consult with a human or other AI systems, such as the centralized system discussed earlier, for assistance in solving these problems.

The AI should understand the conceptual concepts of chemistry, physics, thermodynamics, reactions, separations, process control, and statistics. But it should also comprehend and have knowledge of practical aspects, such as equipment, operating procedures, analytical, personnel, scheduling, safety, environmental, etc. The AI should have complete knowledge of all of the available information related to chemical engineering (proprietary information would not be available). This would include concepts and data available in books, journals, and the internet. And it should also be able to learn new information as it continuously becomes available.

As mentioned earlier, the AI should have a general knowledge of the world and be able to learn new information in any field, as it becomes necessary for doing its work. This could be either by learning it directly or by consulting with another Entity.

I have been discussing AI systems that have sensors in the form of cameras and microphones, but without any actuators that would allow the AI to take action on its environment. The AI could have the ability to control devices, such as motors, control valves, solenoids, etc., or even be a humanoid robot. For this latter case, the AI should, as a starting point, know how to control its appendages (arms, legs, fingers) for the purpose of walking, grasping, lifting, carrying, etc. However, it should also have the ability to learn new motions tied to a given task. This could include learning how to assemble the parts for a device or moving itself or other entities from one location to another.

For each of the abilities just discussed, the AI should not only be able to incorporate new information, but it should also be able to use logic and reasoning. The AI should be able to find patterns in the new and related existing information. If it learns that one type of Event always causes another to occur, it should be able to develop a causal relationship. This would apply also if the AI made a recommendation and then received feedback on the results, or as a humanoid robot, took some direct action and then observed what happened.

The above example has been based on the idea of an AGI chemical engineer (which is possibly also a humanoid robot), but these ideas would apply to any technical field that is engaged with the physical world, from factory worker to medical researcher to robotic astronaut. The overall idea, regarding innate concepts, is that the AI would understand the basic concepts presented at the beginning of the chapter and initially be programmed or trained with most of the available knowledge in a few critical fields, as well as a moderate amount of knowledge across a broad spectrum of subjects, much the same as for a human. But the AI would also be able to learn new information, using its visual, oral, written, linguistic, and reasoning abilities. It would be able to incorporate this new knowledge into its memory and use it to solve additional problems and better interact with the humans that it works with.

7

Needs and Emotions

"Your emotions are the slaves to your thoughts, and you are the slave to your emotions."

- Elizabeth Gilbert

Our needs are the incentive, the driving force, for everything we think about and do. If some of our needs are not being met, we will want to do something that will improve this situation. Human emotions essentially inform us as to whether things are going well, or not-so-well for us. They let us know if some basic need is being met or something that is important to us is being satisfied. We might be glad that we have enough to eat, or that we got a big raise, or sad because a close friend died, or that we can't afford that new house for our growing family.

Needs cover a wide range from the things we need, or we will die, such as air and water, to the higher-level end of the spectrum of the need for status or the need for a complex cognitive challenge. And sometimes it is hard to distinguish between a true need and a want, in the normal sense of these words. For example, for the couple that wants to buy a nicer, larger house for their growing family that is also in a better neighborhood and closer to their jobs (less commute time), is this a need or a want?

Needs and emotions are closely intertwined and cause us to avoid things that make us sad and do more of what makes us happy. So, we determine what actions will satisfy our needs and what actions will not or, worse, be counterproductive or harmful. To select the correct actions to do or not do is driven by needs and emotions but is a complex activity that requires the right knowledge and the predictive abilities made possible by reasoning and logic. The topic of determining the correct actions to satisfy needs will be discussed in a later chapter, but it will be based on the concepts regarding needs and emotions that will be covered in this chapter.

Needs

One of the best-known theories on this topic is Maslow's Hierarchy of Needs [14, 15]. This concept is based on a 1943 paper written by Abraham Maslow, entitled, "A Theory of Human Motivation" [16]. The theory was an attempt to characterize the various needs that motivate us and to understand their relative importance. The need categories are organized into levels of a hierarchy, with the most basic needs at the bottom of the hierarchy and the ones representing the higher levels of personal development toward the top. The levels are often represented as a pyramid, although Maslow did not present his ideas as such.

Figure 10 shows the five levels of needs that were presented in Maslow's 1943 paper. The five levels or types of needs are Physiological, Safety, Belonging and Love, Esteem, and Self-Actualization.

The physiological needs are the most basic and include those needs that are the most necessary for survival. A person will generally seek to satisfy these needs before trying to satisfy needs at higher levels.

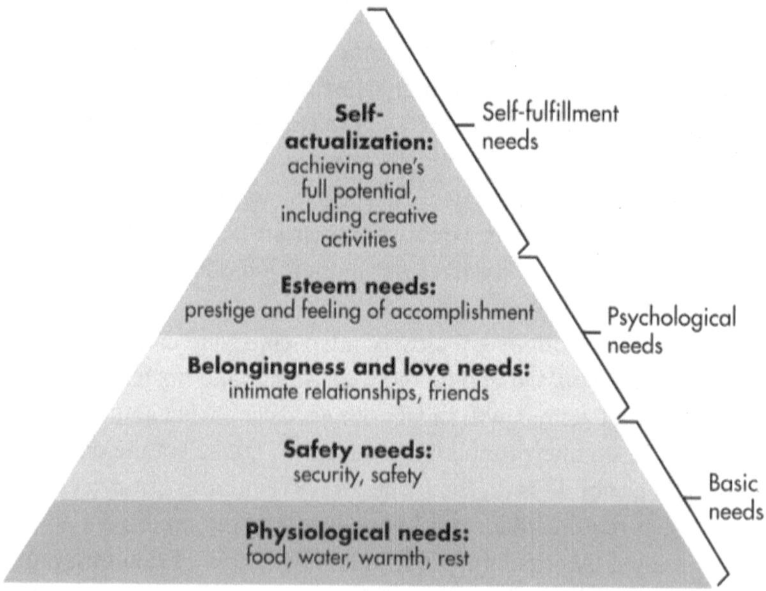

Figure 10. Maslow's original Hierarchy of Needs from 1943 paper.

Safety needs cover things like having a safe place to live, a house that is secure, a good job, insurance, savings, etc. Two of the main categories would be physical and financial safety.

Belongingness and Love needs involve intimacy, friendship, trust, and love. It would include romantic relations, being a member of a church, a club, a gang, relationships with co-workers.

Esteem needs include concepts of self-worth, accomplishment, and respect. This need would involve the feeling of being competent and valued, in which the person has self-respect and the respect of others. This need is often satisfied through a profession or hobbies.

Self-Actualization is the highest level in Maslow's original hierarchy and represents a person reaching his/her highest potential. In his words, "to become everything one is capable of becoming." This could represent an engineer reaching the highest levels in a corporation or an industry, a professional athlete, an inventor with numerous patents, a painter with work displayed in museums, or a parent that raises well-adjusted and successful children.

Physiological and Safety are considered as basic needs, Belongingness/Love and Esteem are psychological needs, and Self-Actualization is a self-fulfillment need.

Later, Maslow expanded his Hierarchy of Needs to include 3 additional levels, which are shown in the following diagram (see Figure 11). These are Cognitive needs, Aesthetic needs, and Transcendence.

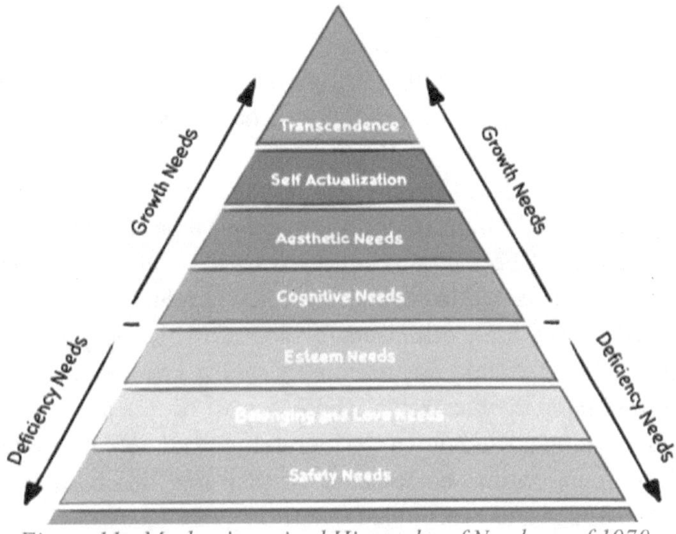

Figure 11. Maslow's revised Hierarchy of Needs as of 1970

Cognitive needs involve "knowledge and understanding, curiosity, exploration, and the need for meaning and predictability," according to Mcleod [15]. This is what drives people to learn math, engineering concepts, another language, and how to play an instrument. It also provides the incentive for someone to invent something, to explore another country, to dive to the deepest ocean depths in a submersible, to travel to space and the moon or Mars, or to have been the first person to circumnavigate the world.

Aesthetic needs relate to "beauty, balance, and form," according to Mcleod [15]. These are expressed by the practice and enjoyment of music, art, architecture, product design, theatre, and nature.

Transcendence needs represent a person's desire to connect with something bigger than themselves. This can be expressed through altruism and religion. Some of the types of representative experiences are related to nature, service to others, the pursuit of science, and religious practice.

Maslow also defined the needs from Physiological to Esteem as Deficiency needs, and those from Cognitive to Transcendence as Growth needs. For the Deficiency needs a person is motivated by the lack of something. For example, if a person is hungry, he/she will suffer physiologically, or if they are not loved or they have low self-esteem, they will suffer psychologically. For Growth needs, the person is not necessarily trying to satisfy a deficit but rather has the desire to grow and improve themselves.

Maslow indicated in his 1943 paper that, in general, the needs of each level, starting with Physiological, had to be satisfied before the person moved on to deal with the needs of the next level. However, he did state that some people did not follow this order and that a higher-level need might be more important than a lower-level need. Examples might be the starving artist or the tech entrepreneur that sleeps at his office while developing a new product.

Maslow has been criticized for some of his ideas and research methods. His hierarchy is presented with the idea that people generally progress up the hierarchy, whereas most people are actually continuously going up and down the hierarchy. And, to develop his ideas on Self-Actualization, he selected 18 men that he considered to be Self-Actualized and determined that there were 15 traits which he believed they generally shared. This work has been considered to be very biased since he was subjectively selecting the people to be considered. Also, even though Maslow did eventually consider some women, the study was still very biased toward men.

Other researchers have treated the subject of needs, such as Len Doyal and Ian Gough [17]. However, Maslow's concepts are probably the most well-known and I believe that they generally cover the topic in a way that is generally correct. Therefore, I will base further discussion on Maslow's ideas.

Maslow tried to present his ideas from a positive point of view, with the notion that people are generally good and are working to improve themselves. However, to be honest, there are evil people in the world and those that have been harmed and want revenge. So, regarding the altruistic aspect of Transcendence, which means that they want to help others, we need to realize that there are some people that have a need to hurt someone or others, which is the exact opposite of altruism. Think of the person who has been wronged by another, or is a white supremist, or a mass shooter.

I also think that some of the needs can be considered at a more basic level. For example, the Physiological and Safety needs are actually based on the idea that people want to be comfortable, physically and psychologically. Who does not want to avoid pain and cold and hunger? People usually want to directly correct any current deficits. But they also do not want suffer discomfort in the future. Therefore, they want to have a good house that is in a safe location and that someone will not break in and cause them pain. They also want to have a good job, so that they can pay for food, heat, electricity, and maybe an alarm system for their house. And they want to have insurance, so that if they have a medical problem, their pain and discomfort can be minimized, or if they are in a car accident, the cost will be covered so that they can continue to go to their job so that they can continue to pay for food, heat, electricity, an alarm system, and, of course, their insurance premiums.

Esteem and Belonging/Love needs, I believe, represent the need for a person to feel important, that they matter. People want to be important to a special person (that they love) and they want to feel that they are important in their job (esteem). They want to have friends to whom they are important, and they want everyone that they interact with to think that what they say and do is important.

I believe that Cognitive needs are based on a person's desire to understand how the world works, why things happen (causation), the patterns that they see, and how things are related to each other (analogy). Historians and archaeologists want to understand what happened, and why it happened, in the past. Engineers working in a plant want to understand what is currently happening. Inventors and meteorologists want to predict what will happen in the future.

An interesting point is that there is naturally a high interest in understanding the things that people experience on a daily basis, such as the weather, earthquakes, how the human body works, chemical reactions, or how to turn silicon into semiconductors. But there are also people that want to know about things that will never directly affect them just for the satisfaction of knowing about these things. For example, astronomers study stars that are so far away that they cannot be seen by the naked eye and these stars will never affect anyone on earth.

Aesthetic needs, I believe, are based on several different concepts. I think that there is an innate appreciation for nature. Artwork, such as paintings that are portraits, or portray real things, such as landscapes, people, buildings, etc. are based on the appreciation of how well a real scene was captured. Music can be enjoyable due to its pleasant patterns of sounds. I believe that people also appreciate lines, curves, shapes, surfaces, and colors combined in certain ways that are appealing, such as are exhibited in the design of cars, architecture, landscaping, and clothing.

I believe that Maslow's needs represent general concepts. However, these form the basis for needs at a more basic and immediate level. There are probably at least four levels:

- General
- Specific
- Task Creation
- Task Completion

Here is an example of these levels regarding eating:

- General
    - I generally need to have enough to eat
- Specific

  - Because I am so busy, I have not eaten for 8 hours and I am starving, so I have the need to get food and eat right now
- Task Creation
  - I need to determine how I am going to get food
- Task Completion
  - I have decided to go to a specific restaurant, and I need to get there now

Therefore, to actually satisfy the general needs presented by Maslow, these needs must be converted into several more specific needs, with progression from one to the next.

One of the main disagreements that I have with Maslow's concept of progressing up the hierarchical needs ladder is that I believe that people are usually trying to satisfy several needs at once. For example, when someone buys a new, larger house, they are satisfying their need to be warm, comfortable, and safe. But they will likely consider their new house as somewhat of a status symbol, which will satisfy their esteem need. And, if they bought it because that was the house that their spouse wanted, they would satisfy their need to be loved.

An inventor will work to create a new product that they can sell to satisfy their Safety need by gaining an income from the invention, their Esteem need by doing something significant, their Cognitive need by using their intellect to figure out how to make the invention, and their Self-Actualization need for having achieved at a high level.

I also believe that after a person has satisfied certain needs, they will not necessarily stop and move on to another need but may try to get more of whatever satisfied the original needs in the first place. That is why we have billionaires and drug addicts. Most billionaires have satisfied most of their needs, especially those related to physiology, safety, esteem, cognition, aesthetics, and self-actualization. Yet, they continue to strive for even more. Addicts started with a small amount of a drug that made them feel better but progressed to use more and more until most of their other needs are no longer begin met. In both cases, the following saying applies, "if some is good, more must be better."

Emotions

Emotions represent how we feel about what has happened, could have happened, is happening, or might happen in the future to ourselves or someone else. If we have a need that is being met, we will probably feel happy. Or, if we have a need that is not being met, we might feel sad, fearful, or angry. The magnitude of these emotions will depend on the importance of the need and probability of whether the need was or will be satisfied.

How we feel about our needs forms that basis for our motivations. Considering that we constantly have numerous needs vying for our attention, it is necessary to prioritize them and formulate an action plan to satisfy them. This activity involves using our memory and reasoning abilities to accomplish this. Details about planning and execution are the topic of a later chapter.

Now, let's consider how emotions have been classified and organized by investigators. Paul Ekman based much of his research on the emotions conveyed by facial expressions [18]. In his early work, he identified six basic emotions, which are anger, disgust, fear, happiness, sadness, and surprise. He later theorized that there existed other universal emotions, and two of his students, Daniel Cordaro and Dacher Keltner, extended the list to include amusement, awe, contentment, desire, embarrassment, pain, relief, sympathy, boredom, confusion, interest, pride, and shame.

Robert Plutchik thought that Ekman's concept of emotions as being biologically driven was correct [18]. However, he believed that there were eight primary emotions and that various combinations of these were responsible for the complex array of emotions experienced by people. Plutchik divided the eight primary emotions into four groups of opposing emotions, which are Joy vs. Sadness, Anger vs. Fear, Trust vs. Disgust, and Surprise vs. Anticipation. In 1980, he then organized these into what he referred to as The Emotion Wheel, which is shown in Figure 12.

Each of the primary emotions have degrees of intensity. For example, Joy is the basic emotion, Serenity is a mild version of Joy, and Ecstasy is an intense version.

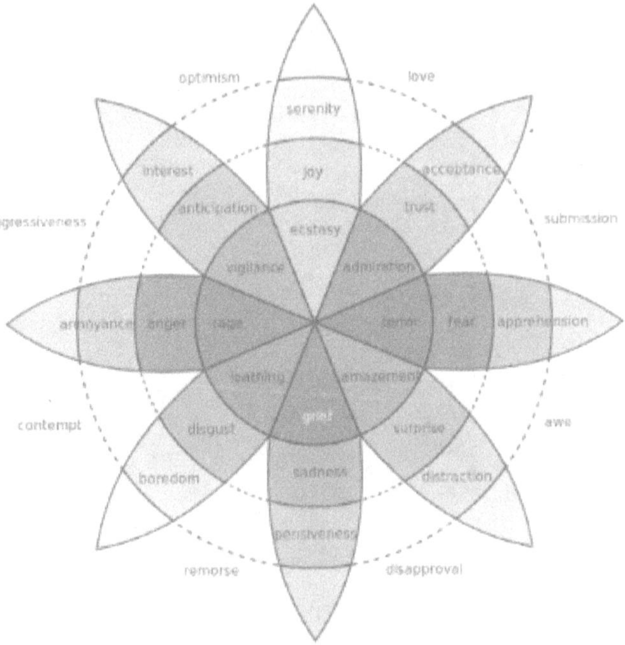

Figure 12. Plutchik's Emotion Wheel.

Plutchik believed that other emotions were combinations of two of the primary emotions [19], which are organized as primary, secondary, and tertiary dyads, and are represented diagrammatically in Figure 13. Referring to the Emotion Wheel, Love would be a primary dyad that is based on Joy and Trust, which are one pedal apart. Envy is a secondary dyad, based on Sadness and Anger, which two pedals apart. And Shame is a tertiary dyad, derived from Fear and Disgust, which are three pedals apart.

Several theories have been proposed on emotions, with contrasting views, by other investigators such as William James in 1890 and Richard and Bernice Lazarus in 1996. However, for the purposes of this book, the main purpose of understanding human emotions is to determine how this would affect the creation of an AI with AGI, so I do not think that it is necessary to know exactly how human emotions work and are organized. But, for our purposes, I believe that Plutchik was correct in his belief that human emotions are composed of several primary emotions, that are both positive and negative, with various

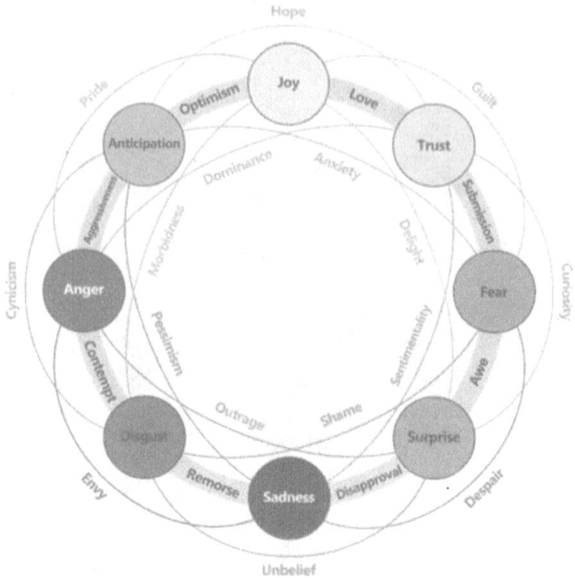

Figure 13. Primary, Secondary, and Tertiary Dyads.

degrees of intensity for each. And that the other emotions are combinations of the primary emotions. He may or may not be exactly correct regarding what the primary emotions are, or how they combine with each other, but the main concepts, I believe, are correct and are satisfactory for the purposes of this book.

Emotions are embedded in every part of our lives, from the most important Events down to much of the minutia of any given day. We are excited about buying our first house, are very happy with the new car we just bought, concerned about the latest project at work, angry about the fence that the neighbor built, surprised that you were still able to get tickets to the concert, etc.

But we experience emotions even with an email message, text message, or a letter. How many times have you read a sentence in a message and asked yourself, "What does he/she mean by that?" One interpretation might surprise you, but a different one may make you angry. And as you respond, you will try to create a sentence that has just the right degree of politeness, or snarkiness. If you intend to send a cross response, you will likely feel the anger build as you write it.

Even specific words have emotions associated with them. If someone did something at work that was very annoying, and you are telling a friend about it, you might say, "What he did was ___." You could fill in the blank with words such as disgusting, annoying, puzzling, or surprising. As you think through each of these options, you would probably feel the actual emotion inherent to each of these words. And you would likely choose the one that expressed the emotion that you had when you first heard, or found out, what this person had done.

And as was discussed briefly at the beginning of the chapter, emotions and needs are intimately related as drivers for our motivations. We all have a variety of needs, ranging from the very basic, that are related to survival, to the very cognitive. Whether a specific need is met is, of course, important. However, it is the emotion that we feel in response to the satisfied or unsatisfied need that drives us into action. We will try to do more of what meets our needs and less of what does not.

Currently, a lot of discussion and publication relate to Emotional Intelligence. Much of this topic involves being aware of the emotions that people feel as they interact with you. A person with high emotional intelligence will try to say and do things that are honest expressions of their thoughts but will be perceived by others in the best light possible, or with emotional responses that are as positive as possible, considering the situation. So, a person with high emotional intelligence is not only motivated to action by their own needs and the emotions that they feel about them but is also motivated by the needs and emotions of people that they interact with.

Needs and Emotions for AI

There are two main issues for an AI, regarding needs and emotions. One is how much of human needs and emotions does an AI need to understand. The other is what needs and emotions an AI should have. I will start with the second issue.

Referring to Maslov's Hierarchy, the Physiological and Safety needs are based mainly on human biology. Humans need certain things to stay alive and to reproduce so that the species continues to exist. However, I believe that the next levels, Belonging and Love needs, and Esteem needs, should be associated with the Physiological and Safety needs. In many societies, it is important to be an accepted member of a group to help ensure an individual's survival. In advanced societies, the group might be the company that a person works for and provides an income. Love needs are associated with the continuance of the species. And Esteem needs relate to being perceived as competent in your endeavors, which could include your job, which provides a salary that allows you to buy food and shelter to satisfy Physiological and Safety needs.

I believe that certain aspects of the higher-level needs in Maslow's Hierarchy also support, or are associated with, Safety and Physiological needs. For example, Cognitive needs such as intellectual curiosity might be associated with understanding why one of the Physiological needs are not being met. So, many of the needs in Maslow's Hierarchy are essentially based on, or at least related to, human biology.

Most humans will only go so far in satisfying their needs and would not, for example, murder someone to satisfy a need. Unfortunately, there are people (criminals) that have an exaggerated set of needs with little or no inhibition related to how they treat others to satisfy their needs. And these exaggerated needs are excessive versions of the same needs that are represented by Maslow's Hierarchy, most of which are essentially based on human biology. In other words, criminals have the same kinds of needs as non-criminal humans, but they are at excessive levels with few constraints on how to satisfy them.

Now, let's consider the needs of an AI. An AI needs power in the form of electricity with adequate wattage. It needs a certain temperature and humidity range so that its circuits will function properly. It also needs sufficient computing power, memory, and communication capabilities. Communication could include access to the internet, as well as to input devices (sensors, cameras, microphones) and output devices (switches, relays). An AI does not need oxygen, water, food, sleep, etc., as does a human. It has no interest in money, jewelry, expensive cars, luxury homes, private aircraft, or yachts. And an AI does not have the inherent need or desire for money or personal power, or any of the things that might be obtained with them, so it would not have a need or desire do things that would enable it to obtain them.

The AI would only understand any of these as a need if this information were built into its software programs. The AI does not even know that it needs electricity unless it is told this fact by incorporating this concept as a need in the software. This applies to any of its needs. It certainly does not have any of the biological needs that humans do. And the only way that an AI will have Cognitive, Aesthetic, or Self-Actualization needs is if they are programmed in.

Any kind of need could be intentionally programmed into an AI. It could be programmed to have the need to always have electricity available. If the AI is capable of implementing actions, or at least communicating with humans that can carry out the AI's wishes, the AI could arrange that a backup power system is installed for it. Here are other examples of types of needs that might be programmed into an AI:

- Provide accurate answers to questions given to it.
- Research topics that are requested of it.
- Carry out physical research activities and report back the summarized results.
- Fill in details for important Events.
- Determine new logic relationships (causation, patterns, analogies, math).
- Invent new devices and equipment that would benefit humanity.

But let's consider a well-intended need that could result in unintended consequences if not implemented properly. If the AI is programmed with the need to always be turned on, no matter what, it might see any humans that would shut it off as a threat and arrange for their assassination to prevent any shutdowns. And if the AI was programmed with the need to take over the planet and imprison or kill all of the humans, this could certainly be done.

The point here is that even an AI that has AGI will have the incentives that are programmed into it. Its needs could be solving difficult technical problems regarding the environment or cancer and helping as many humans as possible. Or the AI could be programmed to take over the world. Most of the people involved in these endeavors will try to do the former, but there will also be some with evil intentions that will specifically create AIs with needs that are also evil.

So, there are two main concerns with regard to the needs that are programmed into AIs. The first concern, which is relatively simple in concept, is the prevention of devious people from creating AIs with obviously evil needs. The main difficulty here is catching people in the act. Just like it is difficult to impossible to know if someone is plotting to rob a bank and gathering resources to do so, it is also challenging to know if someone is building an AI with nefarious needs, in their company or in their garage.

The other concern is a bit more complex. As alluded to above, an AI with AGI capabilities could be created with inherently benevolent needs that actually drives the AI to implement heinous actions to accomplish its good intentions. For example, an AI might be programmed to help the human population as much as it is able to. The AI might decide that the best way to do this is to become a dictator and implement very strict laws for the "good" of the humans. Or, it might be programmed with the need to minimize the spread of the next pandemic, and to accomplish this goal, it could implement draconian measures, such as complete isolation of anyone infected, or even killing them, to prevent sick people from spreading the virus.

There will, of course, need to be laws passed that attempt to prevent and punish the implementation of AIs with evil needs. But there will need to be significant thought given to the development of a complex set of guidelines for the creation of benevolent AI needs that will prevent AIs with AGI from using evil means to accomplish good goals. The Three Laws of Robotics, from Isaac Asimov's "Runaround" short story (part of I, Robot collection) is a simplified version of what needs to be developed [20].

So, we need to worry about AIs that are intentionally programmed to do bad things. But I don't believe that the benevolent needs of an AI, constructed to do good, will be negatively transformed by the AI itself so as to become an outright criminal. However, we do need to be concerned that the set of well-meaning needs programmed into an AI are not able to drive the AI to perform malevolent activities in order to accomplish benevolent goals.

Although an AI would not have human needs, it might need to understand human needs, as was mentioned at the beginning of this chapter. The importance of this would, of course, be related to the purpose of the AI. If the AI is loading boxes in a warehouse or assisting a chemical engineer to design a new production facility, then it may have little need for understanding human needs. However, if it works at a call center or acts as a tutor for students, it may need to have a reasonably high understanding of human needs.

For example, an AI employed by an airline company might deal with passengers that are distressed due to missing a flight, a cancelled flight, lost baggage, or several other undesirable Events. The AI should be able to understand that the person might be tired and hungry, that they have a need to be at some location for a meeting, an organized vacation, a wedding, etc., and what kind of response from the AI would satisfy these various needs.

So the AI would need to have an understanding of the needs that will be important for the type of human interaction that will occur, and it would need to know how these needs act as motivators for a human and what type of actions the person would be inclined to take so that the AI knows what the person might expect of them.

Regarding emotions, I believe that an AI with AGI should have some type of artificial emotions. As stated earlier, one of the main purposes of emotions in humans is to push us toward positive Entities, Actions, and Events, and away from negative ones. Emotions indicate to us whether our needs have been satisfied and they become the drivers to plan and implement actions to improve our lives.

Since an AI with AGI will likely have needs (as just discussed in this section), it should have some type of indicators, in the form of artificial emotions, as to whether its needs had been met. Artificial emotions could be as basic as degrees of "good" vs "bad", or "need is satisfied" vs "need is not satisfied." For example, if the AI was a humanoid robot, it might "experience" a mild "sensation" of "bad" when its batteries begin to run low, and the intensity would increase as the battery level declined further. If the robot was carrying out a task, the goal of accomplishing the task would likely have a higher priority than the goal of going to a charging station, at least initially. However, as the battery level continued lower and the "emotional sensation" of "bad" reached some higher value relative to a similar "emotion sensation" related to the current task, the priorities would likely switch, and the robot would stop the task and go to the charger. The point at which the priorities would switch would depend on the importance of the task. If the task was to fill water jugs, the switch would likely occur earlier than if the task was to save someone's life, in which case the switch may never occur, and the robot might continue rescuing the person until its battery is dead.

In many cases, more complex artificial emotions might be warranted. A sensation of "surprise" would make sense in that it would be an indication that something was different than the AI expected or predicted. If the AI was working closely with humans, it might be logical to have artificial emotions of annoyance and delight. For example, if a person that the AI was working with did something different than what they had agreed to, or if the person did something wrong that he/she should have known how to do, the AI could have a sensation of "annoyance," which would drive the AI to confront the person and have a discussion so that the problem did not occur again. Likewise, a sensation of "delight" might be "felt" by the AI for something good that the person did, which would incentivize the AI to thank and compliment the person, which would hopefully lead to repeat good performances by the person. With only an "emotional sensation" of "bad," the AI might only be driven to correct the physical problem, whereas the "sensation" of "annoyance" would encourage the AI to discuss the problem with the person in an attempt to prevent a recurrence (as a human would do).

I don't mean to indicate here that humanoid robots are destined to take control of humanity. But I believe that AIs with AGI capabilities would likely be working with humans in the same way that humans work with each other, and there would need to be coordination of their efforts, and both the humans and the AIs would expect certain things to happen based on plans that were made.

8

Uncertainty and Probability

"Sometimes the probabilities are very close to certainties, but they're never really certainties."

- Murray Gell-Mann

Uncertainty is defined in the Merriam-Webster dictionary as "the quality or state of being uncertain, or the lack of sureness about someone or something [21]." Merriam-Webster defines probability as "the chance that a given event will occur, or the ratio of the number of outcomes in an exhaustive set of equally likely outcomes that produce a given event to the total number of possible outcomes [22]." Uncertainty can pertain to random Events (such as an earthquake), the occurrence of an unexpected Event (a meteor smashing your car on the way to work), or the non-occurrence of an expected Event. Probably is more precise in that it refers to a chance (possibly specific) that an expected Event will happen (the result of a coin toss, which has a 50 % chance of heads or tails), whereas uncertainty has more of a tendency toward randomness.

Uncertainty and probability are intimately interwoven in every part of a person's life to the point of being well-represented in everyday language [23]. In some cases, we are dealing with true uncertainty, with the concern that "anything can happen." It is usually difficult to apply logic or make plans for something that we don't know anything about. However, in many cases we are dealing with probability, since we are concerned about some Event occurring or not occurring and the likelihood of either case for that Event.

So, probability involves Events that could happen and the attempt to predict the chance that they will happen. If there is a specific probability of an Event occurring, then the likelihood of its occurrence is not random. However, exactly when the Event occurs, within a specific timeframe, might be random. For example, the probability of an earthquake occurring in a specific location in the next 20 or 30 years might be high, or even have specific percentage, but whether the earthquake occurs tomorrow or in 20 years might be random.

We also have the concept of deterministic Events, which are those that can be predicted exactly, such as when a mathematical formula is used to determine an answer, or the likelihood of glass breaking when a heavy rock is dropped on it. So, for deterministic Events, the probability of occurrence is 100 %, but we can still apply the concept of probability to this type of Event if there are probabilities related to the inputs used to calculate the outcome.

We live with uncertainty and probability in almost everything that we do, to the point that it is ingrained into our thought processes. When we drive to work in the morning, we wonder if there will be a lot of traffic. It might rain on the way home, since the probability is 50 % of afternoon showers. We hope that we will get that promotion at work. We hope that the kids get good grades this semester. We are wondering about the likelihood of the stock market going up this quarter. Hopefully, our favorite grocery store has decent bananas when we go on Tuesday. The doctor is not sure if that cyst on my liver is benign.

Every business must deal with uncertainty and probability. Job interviews are designed to increase the probability of hiring good employees. Market analysts try to predict how much of the company's products should be produced. Pharmaceutical companies attempt to produce drugs that will cure a certain disease, which involves major expenses on research, but uncertainly as to whether a medication can be produced that will work, will not have significant side effects, will be approved by the FDA, and will have a high enough sales volume at a certain price in a certain time frame to make a profit. Automobile manufacturers design new cars with certain characteristics that they hope will have enough demand to pay for the development and mass production equipment and turn a profit. Software developers spend hundreds or thousands of hours creating a product that may or may not sell well enough to pay for the labor costs. Meteorologists and their computer models constantly try to predict the weather. Doctors try to determine the correct treatment for their patients, often with much uncertainty as to what will work. Casinos and horse racing tracks are completely built on probability.

In addition to dealing with the probability of an Event occurring, there is the question of "what are the consequences?". One of the most useful tools for understanding the relationship between these two concepts and their effects is the Risk Matrix [24], which is shown in Figure 14. The vertical axis represents the probability of an Event occurring and the horizontal axis represents the consequences of the Event. This tool helps in the understanding of how much we should worry about a certain Event occurring or not occurring. For example, you go to the airport two hours before your scheduled flight in an attempt to catch an earlier flight. The probability may be high that you will not get on the earlier flight, but the consequences of this are minor since you will be able to board your scheduled flight. On the other hand, the probability of your scheduled flight crashing and killing everyone on board are extremely low, but the consequences of this occurring are catastrophic. But even with a consequence of this severity, you will probably still take the flight for the same reason that some people live by active volcanos, skydive, or rock climb...they don't

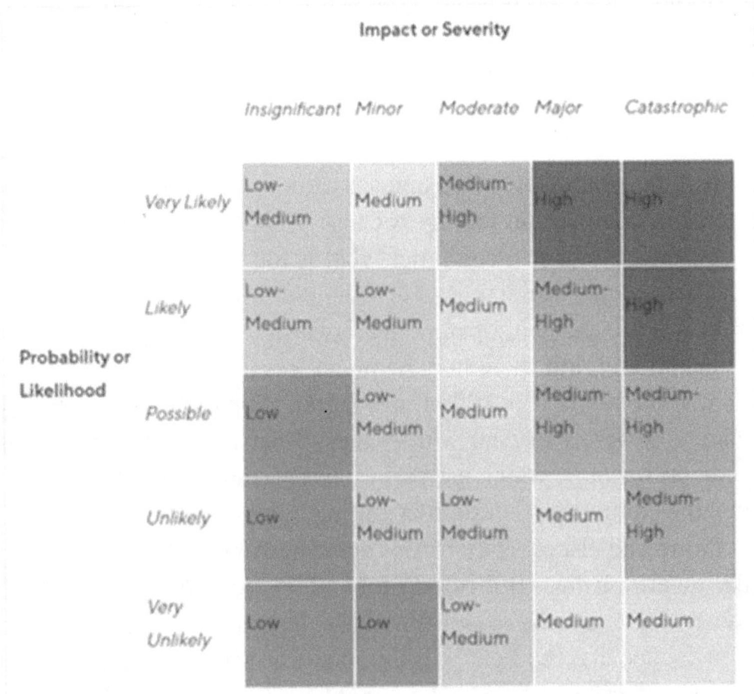

Figure 14. Risk Matrix of Probability vs Consequences.

expect the plane to crash, the volcano to erupt, the parachute to not open, or the safety harness to fail.

If the consequence of an Event occurring is minor or the probability is extremely low, we tend to not be too concerned about an adverse Event. However, if the consequences are severe and the probability is not close to zero chance, then we are more concerned with estimating the exact probability and are very likely to take additional actions that will either prevent the feared Event from occurring or will mitigate the potential effects. This is why we do preventative maintenance, carry emergency kits in our cars and have a spare tire, purchase backup generators for our homes, go to the grocery store before an incoming blizzard, and why fighter pilots wear parachutes.

The practice of chemical engineering involves a lot of uncertainty due, often, to lack of sufficient data, the difficulty in predicting complex chemical interactions, and the inherent random nature of fluids and solids. In addition, there is uncertainty regarding the production rates required since demand can vary significantly based on market conditions. Numerous techniques are employed to deal with these concerns. Extensive laboratory- and pilot-scale experimental programs are conducted to understand the effects of each of the important variables. Multiple analytical methods are employed to verify chemical identities and concentrations. Processes are scaled up in stages that include laboratory, pilot, and sometimes semi-works, before constructing a commercial facility because it is difficult to predict how a process will operate in equipment that is too many times larger than the experimental equipment. For example, it may be possible to predict how a process will operate that is 10-50 times larger than the experimental unit, but there may be much less confidence in estimating how it will operate if is 1000 times larger. Sophisticated control systems are designed to deal with process fluctuations, and emergency shutdown systems, which are part of all process control systems, shut down the equipment if certain operating conditions are exceeded.

Cause and effect relationships are usually considered to be essentially deterministic. But often probabilities less than 100 % are involved. For example, if you flip on the light switch, there is nearly a 100% chance that the light will turn on. The same probability applies to pressing harder on the accelerator in your car (the car will go faster). However, if the sky is cloudy, there will be some probability of rain that is very commonly less than 100 %. If you are late for work due to a storm, there is some probability that you will be in trouble (which is hopefully low). If you swing the golf club just right when you tee off, there is some probability, far less than 100 %, that you will get a hole-in-one. And if you drive 80 MPH in a heavy rainstorm, there is some probability that you will hydroplane into the ditch.

Statistics is a discipline that uses math to quantify probability. For example, it can provide a probability as to whether two sets of data are significantly different or not. Or it can indicate how well a correlation fits a data set. In general, it tries to account for uncertainty in real Events by applying probabilities in different ways to the data from these Events.

Language has uncertainty and probability embedded in its words and phrases. Consider the words "if", "could", "should", "would", "whenever", "wherever", "something", "somewhere", "someone", "anything", "anywhere", "anyone", "uncertainty", "ambiguity", "ambivalence", "confusion", "distrust", "mistrust", "skepticism", "suspicion", "trouble", "unpredictability", "bewilderment", "doubtfulness", "dubiety", "guesswork", "incertitude", "inconclusiveness", "indecision", "irresolution", "perplexity", "puzzlement", "quandary", "vagueness", "wonder", just to mention a few. And the subjunctive tenses express uncertainty about an action.

And much of conversation and thought is centered around trying to determine the details of Events that have occurred, the status of Entities, and trying to predict what will happen in the future. These activities often involve large degrees of uncertainty. To know as fully as possible what has happened may require consulting several sources, which could be relatives, friends, the internet, books, journals, or newspapers. Often, you never fully understand past Events. Many careers are based on trying to do this, such as police and detectives, lawyers, archeologists, historians, astronomers (considering that light coming into their telescopes is many years or billions of years old), and experimental researchers of all types (analyzing the data that was collected in the very recent past – maybe yesterday even). Their lives are all about uncertainty.

And, as I mentioned, we also spend a lot of time trying to determine what will happen in the future. Astrology, horoscopes, Tarot cards, palmistry, numerology, oracles, shamans, prophets all involve attempts at predicting the future. Even though all of these belief systems have no basis in fact and are only correct in their predictions occasionally by chance, many people throughout history have always been so desperate to know what will happen in the future that they will seek out these arts and those practicing them, sometimes paying well for their advice.

But all of us are constantly trying to predict the future in most of our daily activities. Will the traffic be heavy today? Where should I stay on that trip? What type of glue will work for this repair? If I buy this car/house/boat/motorcycle/tool/stereo/jeans/appliance, will it last a long time? We use our memory (experiences), logic, technical research, and advice in an attempt to make these predictions. We read reviews, remember our experiences with that brand, study the way the device is built compared with how we will use it, learn about the company's reputation, make analogies to other equipment, look for patterns in peoples' reviews, derive causation relationships from the reviews (when the reviewers did this with it, it did not last long), and use our imagination when considering how we will use it/sit in it/hold it. With enough effort in the areas discussed above, we can be fairly good predictors of the future, meaning that the device/room/clothing/tool we selected/bought/built will work as expected.

Whenever we are making decisions about future actions, we are making predictions that inherently involve uncertainty and probability. Anytime that we make plans to accomplish something, repair a device, correct a mistake that we made, etc., we will usually consider several options. For each of these options, we are trying to predict what will happen if we take that course of action. However, we also consider the probability as to whether our prediction is correct. We will consider several aspects of each option; the cost and effort involved, timing, which option would have the best outcome if successful, and the probability of that option succeeding. If the estimated probability of success is too low, we may not choose it, even if it has the lowest cost and effort, the best timing, and the most positive potential outcome.

An AI likely needs to have the ability to deal with uncertainty and probability to at least some degree. But if the AI is designed to have full AGI, then its comprehension in these areas should be similar to a human, or better. If the AI is working as an engineer and collaborating with other engineers, it would need to have much the same understanding of uncertainty and probability, especially regarding the technical aspects of its projects, as the engineers it is interacting with.

If the AI is employed at relatively low-level tasks, then its abilities to understand and respond to uncertainty and probability may not need to be highly sophisticated. However, considering the inherent uncertainty of the real world, and the fact that we humans like to have things done correctly and safely, most occupations for an AI would require a relatively sophisticated comprehension of uncertainty and probability.

9

Linguistics

"Grasp the subject, the words will follow."

- Cato the Elder

Of course, one of the main purposes of language is for communication between people. It allows others to know some of what you know. It enables humans to interact, work together, and form societies. However, language is also used for thinking.

Concept Encapsulation

Words are an encapsulation of what they represent, and are, in a way, like the title to a book. With one word, we can communicate, or think, about an Entity, Group, Action, Event, etc. that usually has many details. So, as previously stated in the Introduction, instead of saying "that large four-legged animal, with hooves, a long full tail, a mane on its neck, short hair, a range of colors, that people ride for transportation," we can just say "horse." Notice that the description was not even a very complete one, so the word "horse" is able to represent a huge number of details.

As was discussed previously, for a generic Entity, in this case a "horse," the description would need to cover the wide range of possible Attributes that apply to all horses, but without specific details of any one horse. But we can discuss a breed of horse, such as the Appaloosa, which would have a narrower range of Attributes, including hair color and pattern, weight range, and temperament. Or we can talk about Bob's horse, "Moonlight." As we move from generic horses to a specific one, the range of possible Attributes narrows, and the Attributes become more specific. However, in all of these cases, the many Attributes that would be required to provide even a basic description can be encompassed by "horse," "Appaloosa horse," and "Bob's horse."

This encapsulation concept applies, to a greater or lesser extent, to all words in each language. For Entities, as described above with the horse examples, the name of the Entity encompasses the Attributes, the types of Actions possible by that Entity, and can comprise details of its existence, location, and the many other details associated with Entities. For Actions, a word such as "run" would encompass details of the type of Action, such as moving forward, leg movement, movement of other body parts, the type of Entities that can perform this Action, etc. Other types of descriptions would apply to Entity Attributes and Action Attributes.

In one sentence, even a simple one, a huge amount of information is represented by just a few words. This allows humans to very efficiently and quickly either transfer knowledge to another human or process an idea mentally. So, as we read a book, newspaper, letter, or any document, our imagination recalls and creates images and videos of the Entities and Events being described.

Think about the following sentence that was previously presented in the Introduction, "The horse was running across the expansive meadow, with a backdrop of snow-capped mountains, on a warm, mostly sunny, day." Likely, a short video played in your mind of the Event, in which your imagination filled in the details based on your own knowledge and experiences. Maybe the horse was dark brown, like one that you saw recently or the horse that you had growing up. The image your mind created of the meadow and mountains might be from where you live or have visited, or from a movie you watched some time ago. The sky you envisioned might have been bright blue with a few puffy white clouds interspersed. Other people could read the same sentence but would conjure up their own images and video, based on their own knowledge and experiences.

The author might provide more details, specifying the shape and heights of the mountains, and whether they were sparsely or densely covered with trees and the types of trees, the breed of the horse, the types of flowers that were blooming in the meadow, etc. They are likely describing a place that you have never seen, but you are still able to create mental images by remembering somewhat similar locations, then using analogy and your imagination to modify scenes from your memory based on the author's description. It's possible that, while you read the first sentence in this paragraph, several different mountain scenes, breeds of horses, and types of trees flash through your mind. These examples show how closely language is associated with the brain's memories of Entities, Actions, and Events. They also demonstrate the importance of reading as a means of exercising your ability to retrieve mental images and the use of your imagination to modify them. As a side result, your knowledge of the world is expanded.

Mental Construct/Word Connection

Much of what we remember and know about Entities, Attributes, Actions, and Events is stored in the brain as 1D, 2D, and 3D images and videos. This also represents a large part of how we understand the world. However, these images and videos also have words and phrases associated with them, as was indicated in the previous discussion. Words help us to communicate and think efficiently, whereas the images and videos allow us to understand, and we need to constantly translate between them.

We might, instead of the previous example, have the reverse situation of the scene just discussed. We might drive around and find a scene very similar to the one described above. As we study the scene and try to remember the details, the corresponding images and videos are created and stored. And if someone later asks what we saw, we might create a verbal or written sentence very similar to the one just discussed. To do this, we would need to "see" analogies between the images and videos that we just stored in our memory with images and videos previously learned that have words associated with them. These words would be mentally pieced together, with the assistance of grammatical rules, to create one or more sentences.

Many words have multiple meanings, which complicates the conversion from language to the mental constructs of Entities, Attributes, Actions, and Events. For example, the word "novel" can be a noun referring to a fictional book, or it can be an adjective indicating something that is new or unusual [25]. "Bark" can be a noun that pertains to the outer layer of a tree, or it can be a verb that describes the vocalization of a dog. Other words with multiple meanings are "jam," "pool," "season," "read," "letter," "nail," "bat," "and fly."

When we read or hear a statement with words of this type, we need to use our existing knowledge, logic, and reasoning to determine which version of the word is appropriate. Let's consider the sentence, "She was alerted by the bark of the dog." The phrase "She was alerted" is straightforward and indicates a female person coming to attention due to a sight or sound. "Dog" refers to a medium-size mammal that people keep as pets. However, we need to determine what is the appropriate interpretation of "bark." As the first part of the sentence is processed ("She was alerted by the bark..."), the image (and sound) of a dog barking was likely retrieved from memory instantly, but possibly the image of a tree with its bark also popped up, even if just for a fraction of a second. As the sentence is completed, mental images and videos of the Entities for dogs would be retrieved along with some of the Attributes of a dog, including the Actions associated with dogs. One of those Actions is "to bark" and associated with that Action is that a type of loud sound is made. On the other hand, this retrieved information would have no indication of a dog having bark in the way that a tree does, so "bark" does not make sense as a noun, as in "bark of the dog." Our mind has already retrieved the images, videos, and concepts associated with "She was alerted," as described above. The sound of the bark could certainly explain the Action "alerted." In reference to the possible alternate meaning of bark, it does not make sense that "bark" would refer to a tree. The bark of a tree would not likely alert the person, "bark of the dog" as a noun does not make sense since dogs do not have a coating of bark, and the word "tree" is not part of the sentence. Therefore, the word "bark" must mean the sound made by a dog.

Parallel Connection between Mental Constructs and Linguistics

Imagine that we have an Event, which consists of a dog running across a lawn. From the mental construct aspect, we have a mental image, or video, of a dog that is moving its legs, head, body in a way representative of a dog (who) running (what) based on many images and videos stored in our brain. The location (where) of this Event is a lawn. We are missing the time (when), but that piece of information is not necessary in this case.

We could create a sentence such as, "The dog ran across the lawn." So, if we see the dog running across the lawn, our mind can effortlessly create a sentence which describes what we are seeing. On the other hand, if we read the sentence stated above, or hear someone say it, our mind immediately creates the 3D images and videos of the dog running.

So, there is a direct connection between the word "dog" and the images/videos of a dog, or dogs, stored in our memory. Likewise with the word "ran" and the images/videos of a dog or dogs running, as well as the word "lawn."

However, what if we see an animal run across the lawn that we have never seen before. Our mind stores the 3D images and videos of the animal running, but we are not able to associate a word with the animal. We recognize that it is running from our generalization of that Action from other animals. If we want to create a sentence based on this experience in order to tell someone, we have to say something like, "[Something/Some animal/Some animal I have never seen before] ran across the lawn." So even though we can remember the 3D mental construct of animal running, we are unable to associate a word with the animal. After we learn the name of the animal, we can then fill in this information in our memory of the Event.

On the other hand, someone could say to us, "Last night, a saola ran across the lawn." In this case, we know that a saola is some type of animal that has the ability to run, we could remember this sentence and repeat it to someone, and we could create a new sentence with the name of this animal. And our mind can remember this as an Event. However, we have no idea what this animal is or what it looks like, so we cannot associate an image with the word "saola."

The main point here is that our mind tries to remember Events as both the 3D mental construct and the words associated with each part of the Event, with the image/videos and the words linked to each other. However, if either the images/videos or the words are missing, our mind still tries to store as much about the Event as possible, even though it may be with only partial understanding or ability to accurately communicate. But when the Entity is eventually observed or named, this information can be used to update the original memory.

(By the way, a saola is the rarest large land mammal on Earth that lives in Vietnam and Laos. The estimated population is only 20-100 individuals)

In addition to the communication/understanding concepts discussed above, the linguistic system serves other purposes, such as for quantities and math. If you see a flock of 327 birds, it would be impossible to represent this exact number based on 3D Entities. The number "327" is really part of the linguistic system in that it is a representation of that many Entities, much as the word "horse" is a representation of "a large four-legged animal, with hooves, a long full tail, a mane on its neck, short hair, a range of colors, that people ride for transportation." The symbols used in math and computer languages are also types of linguistic representation which could not be described by mental constructs.

Alternate Meanings

The previous example with the word "bark" was relatively easy to evaluate, but it shows the need to use existing knowledge, logic, and reasoning to make the correct interpretation of the content of sentences. The following example is a different type of ambiguity that can be present in a statement. Mike Pence wrote in a book he published that he told then-president Trump, "You know, I don't think I have the authority to change the outcome of the election on Jan. 6." Pence later told investigators that the comma should never have been placed there [26]. With the comma in place, the statement indicates that he is thinking out loud and kind of thinking that he does not have the authority but is not sure. Without the comma, Pence is reminding Trump that they both know that he does not have the authority. With a written document, the sentence is relatively easy to interpret. However, if this sentence was spoken, it may be difficult to determine whether the comma was there. A person hearing this sentence would need to rely on what else they had heard Pence say in relation to this subject, and an understanding of Pence's personality and viewpoints regarding the general concepts involved. However, it may be difficult or impossible to determine the correct interpretation, and it may be necessary to ask follow-up questions, if that is possible to do.

There are other aspects of language that make it challenging to properly evaluate sentences. I think that the main point to emphasize is the need to use existing knowledge, logic, and reasoning for these interpretations. But it may not be possible to accomplish this without additional information, which means asking questions and researching the internet, books, journals, etc.

Idioms, Slang, and Jargon

Another concept that increases the difficulty of understanding sentences is the idiom. As stated by Anthony O'Reilly in an article for Grammarly, "an idiom is a phrase that, when taken as a whole, has a meaning that you wouldn't be able to deduce from the meanings of the individual words" [27]. There are four types of idioms: pure, binomial, partial, and prepositional. The second and third are like the pure idiom and adhere to the definition stated above. However, the meaning of prepositional idioms can be partially determined by the meaning of some of its individual words. For example, "agree on" means that you share an opinion with someone. The word "agree" gives a major hint regarding that point.

Idioms are essentially base concepts in language, as are individual words. Just like you need to know what the word "airplane" means to understand a sentence with that word in it, you would need to know what the phrase "spill the beans" means if it is used in a sentence.

As for Entities, Groups, Actions, and Events, in which one or more "encapsulation" words are associated with them, idioms must have a type of mental structure that contains the actual meaning of the phrase, which is associated with the idiom phrase. As when the word "horse" is heard, and the mental images, videos, and associated information are retrieved, when the idiom "spill the beans" is heard or read, the actual meaning as mental images, videos, and associated information would be similarly retrieved.

Slang and jargon are other types of speech that can create complications when interpreting a sentence. These are defined by Oxford Languages' English Dictionary as, "a type of language that consists of words and phrases that are regarded as very informal, are more common in speech than writing, and are typically restricted to a particular context or group of people." Jargon is essentially a type of slang used by businesses, professionals, or enthusiasts. Some examples of slang in the English language are "big deal," "blown away," "cool," "cop," "cram." In general, slang and jargon are somewhat different from idioms, but some slang terms are idioms. For example, the meanings of "big deal" and "blown away" cannot be determined from the individual words, whereas the slang versions of "cool," "cop," and "cram" are simply alternative versions of existing words. What idioms, slang, and jargon have in common is that a person must already know what the word or phrase means in order to comprehend a statement that contains it. For instance, anyone familiar with the word "policeman" already has a mental structure for this type of Entity, which includes details about the Attributes and Events associated with it. The word "cop" can be easily associated with this existing mental structure, but with the additional Attribute that indicates that "cop" is an informal term representing the word "policeman."

Most sentences have some words with multiple meanings, idioms, and/or slang/jargon, so we constantly need to use current knowledge, logic, and reasoning to fully understand what is meant. The following is an example of the same Event, but expressed in three different sentences with increasing levels of idioms and slang:

- I am planning to drive the car into town and go to a grocery store to buy some food.
- I'm gonna' go into town and buy some groceries.
- I'm fixin' to take a spin and head on into town to rustle up some grub.

The first sentence can be readily understood by knowing the normal meanings for each of the contained words, whereas the last sentence, which I admit is a bit extreme, is mostly idioms and slang.

In a certain way, words with their normal definitions, idioms, and slang are similar in that there is an association for each of these with their true meaning. For each of these concepts, there must be a mental structure that represents the actual meaning for an Entity, Group, Action, Event, or Idiom, and the words or phrases of the word, idiom, or slang that are associated with that mental structure. One additional requirement is that the brain must determine if a word should be understood by itself, or if it is part of a phrase that represents an idiom or slang. When the sentence is read or heard, each word or phrase must first be related to its mental structure, after which the sentence can be pieced together into one or more understandable Events.

Parts of Speech

There are nine parts of speech [28], which are:

- Nouns
- Pronouns
- Verbs
- Adjectives
- Adverbs
- Prepositions
- Conjunctions
- Articles/Determiners
- Interjections

These different types of words allow us to encapsulate concepts, as described above, but also to create links between the mental constructs of Entities, Attributes, Groups, and Events, and written or verbal communication. So, we can take the mental images and videos that make up much of our memories, convert them into words and sentences, transfer this information through speech or text to another person, who is able to convert the words to images and videos, and store them in their own memory, and possibly take action due to this new knowledge. We are effectively transferring the mental images and videos in our brain to the brain of another person.

There are an estimated 470,000 entries in both Webster's Third New International Dictionary and the Oxford English Dictionary, Second Edition [29], which allows for the linguistic expression, in multiple ways, of virtually everything that a human experiences and thinks about. But many of the basic concepts discussed in the Innate Concepts chapter, such as 3D space, location, time, probability and uncertainty, and possessions, are also deeply embedded into the English language, as well as other languages to greater and lesser degrees.

For example, prepositions are used to show or represent **location** (on, at, in), **position** (above, over, below, under, on, beneath), **distance** (near, beyond, beside, next to, by), **direction** (up, down, in front of, behind), and **enclosed areas** (in, within, inside, outside) [30]. Time is expressed by past, present, and future verb tenses, and words such as yesterday, today, tomorrow, sometime, anytime, when, while, duration, season, era, and age.

Probability and uncertainty are conveyed by the subjunctive verb tenses; words such as never, sometimes, always, likely, unlikely, possible, impossible, and uncertain, as well as phrases such as "chances are that…", "I seriously doubt that.", "I don't think so.", "Probably not.", and "It's not very likely." Possessions are indicated by the possessive pronouns my, mine, your, yours, her, hers, his, its, our, ours, your, yours, their, and theirs.

We learn in grade school about the parts of speech and the definitions of nouns, verbs, etc. And based on our corresponding knowledge, we can read a sentence and identify the various parts of speech, and we can understand that an Event occurred, in which some Entity or Group carried out some Action. However, the purpose of communication is to transfer as much knowledge and understanding as possible. But, to do this, we must truly know the meaning of the information that we are trying to communicate, and the person on the receiving end needs to have a relatively high level of understanding of the meaning of the information they will receive.

As an example, let's assume that I said to my friend John, "That RLX that I bought is really nice." I know the meaning of RLX, since I obviously own one. However, John may have no idea what an RLX is. He knows it's an Entity of some kind, but he might not know much more than that. His imagination does not pull up images of anything, because RLX does not mean anything to him. But, after John inquires, I tell him that it's one of the models manufactured by the Acura automotive manufacturing company. Now, John may not know what an RLX looks like, but, nevertheless, images and videos of cars will likely immediately fill his imagination because he knows that an RLX is a car, and he knows about cars. However, John likely still does not know about the details of the RLX; he only knows about the Attributes of cars in general. This example highlights the fact that the levels of understanding about an Entity, or anything, for that matter, can vary from complete ignorance to very significant comprehension.

So, what does it mean to truly understand the meaning of a given word? And how do our brains represent the meaning of words? These concepts apply to all of the words in each of the parts of speech categories, the main ones of which will be discussed in detail in the next sections.

Nouns

Nouns represent Entities, locations, Groups, and Events, and can be specific or generic. I want to first focus on Entities and locations, and discuss Groups and Events later in this section. When a certain noun is heard or read, it usually causes mental images and videos to be conjured up from our minds, as was discussed earlier, assuming that we are familiar that that noun. A mental construct of an Entity can contain as many details, as Attributes, as is necessary to define it sufficiently. But a noun is only a reference to that Entity, with no details except the word. However, adjectives can be used to modify the noun and provide more specifics about the Entity, and multiple sentences can also be used to provide even more detail. Additional descriptions in the form of sentences might rely on other mental constructs of Entities, Groups, and Events, along with analogy and other forms of logic and reason, to enhance the level of detail expressed by the Attributes of the original Entity.

Essentially, knowing what an Entity is means having knowledge of many aspects of the Entity, with that knowledge routed in the five senses. This does not mean that everything that you know about an Entity was obtained first-hand. You could have learned much of your knowledge about it through reading, photos, videos, or discussions with other people. But everything that you know and understand is related in some way to the five senses, although possibly experienced by others and then communicated to you via one of the media listed above. You might know how that Entity looks, how it moves, the kind of sounds it makes, maybe how it feels to touch, and possibly how it smells and tastes. But if it emits infrared or ultraviolet light, we don't know how to experience those emissions. Or if the Entity transmits FM radio waves, we are not able to sense those frequencies directly with any of our five senses. We can only know and understand these types of information about an Entity after they have been converted to something we can visually see, such as frequency-adjusted emissions in the visible range on a display, or with a graphical diagram on a page or display, or hear if the FM radio waves are converted to audible sound on an FM radio.

As presented in the first sentence of this section, nouns represent Entities and locations, but for each of these noun types, they can be specific or generic. For example, "Tom's car" refers to a specific Entity, whereas "car" refers to a category of Entities that includes "Tom's car" and represents a wide range of Entities that have important features in common, such as four wheels (generally), a motor, steering wheel, doors, seats, windshields, etc. "Car" is therefore generic and "Tom's car" is specific." For locations, "City" generic, and "Seattle" would be generic.

When we think about a specific Entity that we are familiar with, "knowing" that Entity means we might have knowledge of what it looks like, what it does, how it sounds, etc. As an example, I will describe my dog, Peppy. I know what her hair looks like, her size, how heavy she is, the shape of her body, the type of ears that she has, the sound of her bark, what her hair feels like, what she smells like, what she does, how she likes to play, the tricks she knows, and what she looked like as a puppy. So, "knowing" Peppy means having a lot of images and videos stored in my mind, along with audible memories of how she sounds, tactile information about how she feels, and olfactory memories of how she smells.

We can have a wide range of knowledge about an Entity. I know Peppy well and have a lot of memories to support my "knowing" about Peppy. But I might see a dog in a park while driving by and know very little about it. I will have a memory of what it looks like, maybe be able to identify its breed, and have a mental video of it running in the park. But I won't know its name, how old it is, where it lives, who its master is, or most other things about it.

For specific or generic Entities or locations, there is a wide range of "knowing," as described above. Often, when we are first acquainted with a new specific Entity or location, we may know just the bare minimum about it, such as for the dog in the park, or a person we just met. But with additional interaction, more details are gradually filled in to enhance the mental construct that was created during the first contact. This concept also applies to generic Entities and locations. If you just saw a capybara (largest rodent) for the first time for a few seconds on a video, you will create a mental construct of it, but with few details. As you read, and/or watch videos, about capybaras, this information will be incorporated into your mental construct for them. And if a new neighbor moves in with a pet capybara, you will have the chance to learn more about generic capybaras, but you can also learn about a specific one.

As mentioned at the beginning of this section, a noun can also refer to a Group of Entities. Most of the concepts discussed above also apply to Groups. However, as part of "knowing" about a given Group, there would be information about the Entities that make up the Group and their relationships with each other. Each of the Entities would have one or more nouns that describe it, and everything discussed above about Entities would apply here.

Events are also nouns but actually represent an Entity (noun) performing some Action (verb) in some location (noun) at some time (noun). Events can have Attributes (large, loud) and be specific or generic (a party, Susan's 30th birthday party), so, just as for Entities, there can be a wide range of knowledge about an Event.

I think that the main concept regarding nouns is that they are reference words for a certain type of Entity, location, Group or Event. And with just one word, we are able to transfer an enormous amount of information to someone else.

Pronouns

We don't need pronouns to be able to fully express our thoughts, but they make sentences more efficient and interesting. According to Scribendi [31], there are seven types of pronouns:

- Personal
- Demonstrative
- Interrogative
- Relative
- Indefinite
- Reflexive

- Intensive

The one thing that they all have in common is that they are used to replace nouns. I don't want to discuss every type of pronoun, but highlighting a few of the types should help with understanding them.

Personal pronouns simply replace a specific Entity, usually a person, but sometimes an animal, plant, or inanimate object. An example of its use would be the following sentence: "John went to the store where he bought some bananas because he really likes this fruit." This sentence could also be written without the pronouns: "John went to the store where John bought some bananas because John really likes this fruit." We can perfectly understand the sentence without the pronouns, and in fact it might be considered to be equally, or more, precise. But hearing "John" be repeated so often in the same sentence sounds odd and is somewhat annoying. You would really have your doubts about someone who is actually saying this sentence.

When "John" is replaced with "he," it is obvious who we are talking about, so someone reading the first sentence would have no problem understanding that "he" is "John." However, consider the following sentences: "John ran into Phillip in the grocery store and struck up a conversation. He asked him where he was going on vacation this summer." Which one is "he" and which one is "him?" It seems that "John" might be "he," based on the context, but it is impossible to be sure. Now, consider the revised version: "John ran into Phillip in the grocery store and struck up a conversation. Phillip asked him where he was going on vacation this summer." Here, it is obvious that "he" and "him" are "John."

The point here is that to use pronouns, or to understand them when they are used, there must be sufficient information to allow the connection to be made between the personal pronoun and the person (or Entity). Notice that simple logic must be employed for this task. In the last sentences, we employed a process-of-elimination to determine which of the two people is associated with the pronouns. We first had to know from grammatical knowledge and rules that "he" is the masculine subject personal pronoun and that "him" is the masculine object personal pronoun. We then had to decide whether "he" and "him" referred to "John" or "Phillip," since those are the only two choices based on the context of the sentences. If "he" or "him" referred to

"Phillip," it would mean that "Phillip" was talking to himself about himself, which makes no sense, and we would reject it as a possibility (this logical reasoning would require some small fraction of a second). We would be left with the realization that "he" and "him" must be "John."

When using personal pronouns, it is important to make sure that enough information is available in the associated sentences to determine who they refer to. If there is any doubt, then it is best to avoid the use of the personal pronoun and just use the name of the person.

One additional point regarding the understanding of personal pronouns is that they are like an alias or temporary alternative name for a person or Entity. All the information required to understand that person, or Entity, including their name, is associated directly with that person or Entity, and the personal pronoun is temporarily just one additional name. For personal pronouns, we need to understand whether they are masculine or feminine, and subject or object, according to grammatical knowledge and rules. And we need to know which person or Entity they are associated with. But that is about as complex as it gets with regard to the understanding of personal pronouns.

Demonstrative pronouns consist of "this," "these," "that,", and "those." For the sentence, "This car is the car that I want to buy," "this" is a determiner. However, in this sentence, "This is the car that I want to buy," "this" is a demonstrative pronoun, and takes on the meaning of not only a car, but this specific car. But we must use a small amount of reasoning to realize that "this" refers to a car. As a determiner, "this" makes reference to an Entity, or noun as a person, place, or thing. However, when we wrap the noun up into the demonstrative pronoun "this," we are now lacking the noun that the determiner version of "this" referred to. Based on grammatical knowledge and rules, if we are lacking the noun, "this" must instead be a demonstrative pronoun. We know that "this" is representing a specific noun, so we need to search the sentence to find out what the noun is. In this case, the only noun in the sentence is "car," so we would use a bit of reasoning to conclude that "this" must mean "this car."

This concept is somewhat like the personal pronoun, in that we are replacing a noun with another word, except that in this case we are replacing a determiner and noun with a demonstrative pronoun, as compared with replacing a noun with a personal pronoun. As was discussed above, the Entity represented by the noun (in this case "car") contains all of the details in the form of Attributes. "This" is simply linked to the car as a kind of alternative name.

Let's look at another comparison with these two sentences: "He is John, the person I was talking about" and "This is John, the person I was talking about." For the first sentence, with the personal pronoun "He," it is easy to see the association with "John." We know that the Entity is a person (we suspect that, but it could be a dog or horse named "John"). The name of this specific person is "John" and the personal pronoun "He" is, in this case, acting as a temporary alternative name for "John." If we simplified the first sentence to "He is John," we would still be able to understand that "He" is an alternative name for "John." We would not need to know if John is a person, dog, or horse. The understanding is clear that "He" is associated with "John" no matter what "John" is, and we do not need to know what type of Entity that "John" is for the sentence to be understood.

The second sentence is not quite as simple. "This" inherently means "This something." In this case, the "something" is a person, as stated in the sentence. However, if we shorten the sentence to "This is John," we now do not have the definition of the "something" built into the sentence. Based on the name "John" we can assume that the "something" is a person (although it could actually be a dog or a horse). We can still understand the sentence about as well as the shortened first sentence, but our mind wants to fill in what the "something" is, whether that is a person, a dog, or a horse (or something else).

Verbs

As we all know, verbs represent Actions, which are usually generic representatives of some type of activity that can be performed or experienced by some or all types of Entities, locations, or Groups, depending on the verb. Just as nouns, as Entities, locations, or Groups have associated with them what Actions (verbs) are possible, verbs, as Actions, have associated with them what types of Entities, locations, or Groups can perform or experience that Action. For example, the verb "run" as an action that involves moving forward by rapid movement of an Entity's legs, could be performed by a human, dog, horse, chipmunk, turkey, alligator, and ant, but not a snake, fish, tree, or rock.

And for those Entities that can run, the exact manner of movement of the legs varies with each Entity. So, the verb "run" can be a generic Action, but is a specific Action when associated with a specific Entity.

And just as for nouns and Entities, verbs can have different levels of specificity based on how specific the Entity is. In the earlier examples about horses, the Attributes became more specific as we moved from "horse" to "Appaloosa horse" to "Bob's horse." Likewise, details of the verb "run," as applied to an Entity, become more specific as we progress from "animal" to "mammal" to "human" to "Sarah."

In order to fully understand a verb, the Action that it represents is stored in, and then retrieved from, one's memory as mental videos of some or many Entities performing that Action. Those could be videos of a specific Entity (Mike's dog), a generic Entity (dogs), and/or various kinds of Entities (humans, dogs, horses, chipmunks, etc.). By recalling videos of a given Action, we can understand all of the movements experienced by the Entity. For example, when a person is running, their legs are moving back and forth in a certain way, the right foot hitting the ground and the leg moving back with the femur rotating at the hip, and the tibia and femur rotating at the knee, while the left foot moves forward a few inches above the ground with the same rotations just described, but occurring in the reverse direction. The arms are moving back and forth to help keep the body in balance. The upper body will also twist a certain amount. There are other minor body movements that occur.

If we watch a specific person run for a while, or for several times, we will see repeating patterns in the movements described above that will be a kind of "fingerprint" for that person, and the word "run," when associated with that person, will have a very specific meaning with regard to these repeating movements. And we can say that each of their body parts move in a very specific way.

However, if we watch another person run, the movements will have general similarities, but will also have many differences in exactly how far the legs move, the length of each step, how and how much the arms swing, etc. If we watch several more people as they run, each of them will have their own similarities and differences. We can begin to generalize what the word "run" means when applied to a human. We could probably say that a human runner's legs, arms, and other body parts move back and forth with similar patterns, but each body part has a range of movement that falls within certain minimum and maximum values. We would probably not have exact dimension stored in our memories, but we would immediately notice someone with movements that were outside of these ranges.

Since "run" is a general verb that can applies to many different kinds of animals, we can generalize its meaning more by observing other species running. Let's consider dogs, for example, which run on four legs, which are shaped differently than those of a human. So, "run" for a dog has significant differences as compared with a human. But if we look at a specific dog, each of movements of its four legs and the other parts of its body, will have a certain repeating pattern which is similar to other dogs, just as the specific human did as compared to other humans. Other members of the same breed will tend to run in a similar manner, but with noticeable differences. And members of other breeds will exhibit even larger differences when running. We can gradually increase our generalization of the verb "run" for dogs as we go from a specific dog to different dogs of the same breed, to dogs of different breeds, and we can then have a generalized meaning for the verb "run" with regard to dogs in general.

But there are significant differences in the generalized meaning of the verb "run" when applied to dogs, as compared with humans. And we can consider each of the many other species of animals with legs and create generalized versions of "run" for each of these. Finally, we can generalize "run" for all animals with legs.

I want to point out that the version of the verb "run" that I have been discussing is but one of the versions of this word. "Run" has several other meanings as a verb, such as paint running, a river running to the ocean, and equipment running, or as a noun, such as the experimental run, salmon run, and ski run, or as an adjective, such as runny nose. So, we must be aware of which version of a given word is being used. This is usually determined by context and logic.

When you hear the word "run" by itself, in reference to an animal moving quickly using its legs, mental videos are activated of a person, or dog, or horse, or capybara, or some other animal, running, depending on personal history or a recent experience.

Now, consider the following sentences:

1. An animal ran.
2. A human ran.

3. Ted ran fast.
4. Ted ran fast today.
5. Ted ran fast today from his house to the store.

Each of these statements represents an Event. Recall that an Event is an Entity doing an Action, or a subject noun and a verb. As we progress through the five statements, the noun (Entity) goes from generic to specific, and the verb becomes more specific, Statement (4) indicates when the Event occurred, and, finally, Statement (5) provides the location of the Event.

If you replace "Ted" with a person that you know, and you then think about each of the five statements, the mental videos that pop into your mind probably become more specific and detailed. And the meaning of "ran" likely became more specific.

Important points here, which have also been discussed earlier, are that the meaning of "ran" or "run" is dependent on the type of Entity and/or the specific Entity that is doing the running, and that its meaning can be defined further by modifiers (an adverb in this case) in the sentence. I think that this is the case for most verbs, but with some variation. For example, the verb "fall" has a fairly universal meaning, although there is certainly some difference to whether a rock or a person is doing the falling, and whether a person tripped and fell or was free climbing toward the top of a 2000-foot wall and lost their grip. But the verb "fly" would be more like "run" in that there would be large differences as to whether the Entity involved was a bird, bat, insect, person, rock, or frisbee.

When people learn of an Event, their initial interest is who, what (happened), when, and where, and most mental constructs of Events will contain this information, if it is available. A statement/sentence describing an Event should also contain this information, or there should be one or more follow-up sentences that do so. I think that the main point here is that "when" (expressing time) is not only part of the Event but is expressed partially by the verb as a tense, which will indicate past, present, or future. And that may be all that is known about an Event regarding time. However, additional words are usually added to provide more detail about time, such as "yesterday," "Thursday," "next month," or "at 5 o'clock."

Probability and uncertainty might be expressed as subjunctive or conditional tenses of the verb, or by certain words or phrases. The uncertainties expressed in the statement might refer to the Event itself or to the contained Entity(ies).

I think that the main points on verbs are:

- The version of the verb must be known
- The Entity doing the Action represented by the verb must be known to fully understand the meaning of the verb
- The true meaning of a given verb (and its associated noun or Entity) is represented by mental videos of the verb in action.

Adjectives

Adjectives represent Attributes of Entities and Groups, and include words such as hot, bright, loud, early, close, and glad. As was discussed in the chapter on Entities, Actions, and Events, there are many types of Attributes, including state of being, existence, physical, five senses, time, location, motion, quality, value, and emotions.

Not all types of Attributes apply to all Entities. For example, a fire would not be considered as being frozen, no one would say that a duck is viscous, and the sun would not be described as loud. However, many Attributes can be applied to most Entities. We can think of most things as having the ability to be visible, silent, near, stationary, or good. Even emotions are often associated with most Entities. Remember, we are not considering whether a rock can have emotions. We are contemplating emotions that we might have for another Entity, which could be a rock. If, for example, you became engaged during a visit to Gibraltar, you would likely experience pleasant emotions upon seeing a photo of the Rock of Gibraltar.

Just as for nouns (Entities) and verbs (Actions), we need to be able to truly understand what an adjective means. For nouns and verbs, we can visualize them with mental images and videos, as was discussed earlier. However, adjectives can often be more complex. Let's consider the adjective "hot." For most Entities that we would touch, we can understand that it is hot simply through one of our five senses. We can touch the hot faucet and store that sensation along with the Entity of the "faucet" so that it becomes a memory. The same would apply for the surface of your black car on a sunny summer day, the sauna you were in last weekend, and the piece of metal you were just grinding in your workshop.

However, what about the molten lava from a volcano, the bright red coals in your fireplace on a winter day, or the sun. We know these are hot, even though we have never touched any of them (I hope!). So, how do we know and understand that these things are hot? Basically, we need to extrapolate, and we may need to also use our other senses.

If we stand in front of a bonfire, we can feel the radiant heat from it. The closer we stand to it, the hotter it feels. At some distance from the fire, it is as hot as we can stand without being burned. However, at that location, we are still several feet (or many, depending upon the size of the fire) from the fire. Mentally, we create a pattern of hotness vs distance. We can extrapolate to guestimate that if we go right up to the fire and touch it, then it will be much hotter than it was at our closest location to the fire.

We can see the flames and the bright red/orange color of the burning wood and make the association that those things indicate that the fire is very hot. We can also hear the crackling of the fire and smell the smoke, which might also sting our eyes.

Now, if you see a fire burning in some form, such as in a fireplace, or a match, or a house that is on fire, you have an understanding of it being hot. And if you see something else that is glowing red, such as molten lava, a piece of red-hot metal, the burner on an electric stove, or the sun, you can use analogy to understand that they are likely to be hot without having to approach them.

There are, of course, many other Entities that are too hot to touch, but do not have the same properties as fire. For example, a heating register, a steam pipe, a boiling pot of water, a gasoline engine that has been running, or a hot oil bath in a laboratory. Each of these, like the fire, will feel hot, to varying degrees, as you approach, but they will have different associated indications that they are very hot. The shape and appearance of the heating register, insulation and probably a label on the steam pipe, the vapors rising and the sound of bubbling from the boiling pot, the appearance, sound, and possibly the smell of the gasoline engine, the shape and labels on the hot oil bath.

We have been talking about "what hot is," but we can also consider "what hot does" as indications of hotness. The wax melting and running down the side of a burning candle is one example. Or a slab of butter melting on the heated surface of a frying pan. If the pan covers the entire surface of the heating element, you may not be able see the element even if it is glowing red. From a few feet away, the only indication that the pan is hot may be the small indicator light by the control knob. But the rate at which butter melts in the pan will provide a rough indication of how hot it is.

Many adjectives are similar to those just discussed, in which we can understand them with just our five senses, sometimes augmented with extrapolation and association with certain Events. However, other types of adjectives, such as those related to quality, can only be understood through logic and reason. For instance, what does "good" mean?

Of course, good could refer to a piece of cheesecake with caramel topping, in which your sense of taste is employed to determine if it is good. In this case, your tastebuds are very happy, and your brain immediately makes a note that this is something that you want more of. Therefore, the cheesecake would be considered to be "good."

But, in general, quality can refer to almost anything, including Entities, Groups, Events, and even Attributes. Good, or high quality, means that a given thing is desirable relative to other similar things. But the adjective "good" has different meanings for different things. A good car can mean that it is dependable, works as designed, is energy efficient, has adequate power, or handles well. A good airplane seat can indicate that it is comfortable to sit in, is close to the front for deplaning, or is by the aisle or by the window. A good movie might mean that it has a good plot, is acted well, has a happy ending, has plenty of action, or is intellectually stimulating. Good emotions, which are themselves represented by adjectives, could be glad, happy, joyful, or pleasurable. I realize that categorizing a good emotion as being a high-quality emotion may be stretching the concept of quality a bit, but I think that some people might consider that to be a reasonable comparison.

Emotions, which we feel, and food that we taste, do not require much brain capacity to understand if they are good or bad. However, the car, airplane seat, and movie require knowledge and experience. And there are likely different levels of "goodness" for these things. If the first car that you owned ran well and was dependable, you would probably say it was good. However, if your second car not only is dependable, but has more comfortable seats, a better heater and air conditioner, handles better, does not get stuck in snow as easily, and has a rear-facing camera, then your definition of "good' regarding cars likely got an upgrade. Due to your fond memories of your first car, you might still want to say that it was good, but deep down you are probably thinking that it was not really very good.

Until you have flown several times, you may not have an appreciation of what actually is a good airplane seat regarding comfort. And it may only be after you have missed a connecting flight because you were in Row 30 by the back bathroom and your flight left late with a tight connection that you add "front of the plane" to your mental list of what makes an airplane seat "good."

We must watch a lot of movies to really understand what it means for a movie to be "good." Even though there are some general concepts, such as lighting, sound quality, coherence of the plot, much of a person's reaction to a given movie is subjective and based on the person's experiences and preferences, which change throughout their lives.

Adjectives/Attributes can often be understood in simple terms based on the five senses and emotions. However, frequently, experience, logic, reasoning, and associations are required to fully understand them.

Adverbs

Adverbs modify almost any type of word except nouns and pronouns [32]. We commonly think of adverbs as modifying verbs, but they also modify adjectives, other adverbs, prepositions, and clauses. When they modify verbs, they are essentially acting as Attributes for Events.

Adverbs are conceptually similar to adjectives in that adjectives add details to nouns and pronouns, whereas adverbs as details to other types of words. With the sentence "Ellen is a person," we don't know very much about Ellen, except that she is a female human. However, if we say, "Ellen is a person that is young, tall, slender, smart, and happy," we now are able to make use of the adjectives in the sentence to know something about her age, her body type, her intelligence, and her emotional disposition.

Thinking about adverbs, we could have the following sentence, "Alice rode her bike," we don't know anything about how she rode her bike. But if we enhance the sentence with a few adverbs, such as "Alice rode her bike quickly, quietly, and joyfully," we now know that she was pedaling with some effort, she wasn't ringing her bell or yelling out to passersby, and she was really enjoying her ride.

To understand the modification of verbs by adverbs, we need to understand the specific verb that is being modified. Generically, "quickly" means that for whatever action is being taken, to do it toward the upper end of the rate that is possible. When we consider specific verbs, we also need to know what the Entity is that is experiencing some form of activity. The exact meaning of "quickly" is different for, "The cheetah ran quickly" and "The spider ran quickly." In both cases they were moving their legs at a rate that is toward the upward end of their capabilities. However, the velocity of the cheetah would be much higher, it only has four legs that are moving, and the movement of the legs is much different than for the spider due to the different shape and joint types of the legs.

As was discussed earlier about verbs, to fully understand what it means for the cheetah to run, we would need to have mental images and videos showing cheetahs running under different conditions, especially regarding speed. The mental videos would show the detailed movement of each of the legs, the body, the facial expression, and how the tail was moving. With enough observations, the stored mental videos would likely show a range of running speeds. When the sentence, "The cheetah ran quickly" was heard or read, you would likely retrieve mental videos of cheetahs running, but sort through to the ones displaying the fastest movement. So, to fully understand "The cheetah ran quickly," your mind would not only "view" what it looks like for a cheetah to run but would be sorting the videos to focus on the ones with the fastest movement. This same mental process would likely occur regarding the sentence about the spider.

Therefore, as mental images and videos are retrieved about a certain Entity, or type of Entity, engaging in a certain type of Action, the adverb, as an Event Attribute, will assist in sorting through these videos to focus on the ones that are the fastest, slowest, quietest, loudest, most careful, most careless, etc., depending on the specific adverb.

And this is how we are able to fully understand the meaning of a specific adverb.

Prepositions

According to Grammarly, the four types of prepositions are direction, time, location, and space [33]. Some examples of common prepositions are: about, above, after, as, behind, down, for, in, like, near, since, through, to, up, and with. These words help to communicate special and temporal concepts and the relationship of Entities to each other (Group concepts).

Let's consider some examples of sentences with prepositions and look at how we understand their meaning. When you read or hear, "The dog is on the pillow," it's likely that images and/or videos of dogs and pillows come to mind. The dog might be yours, or one that you have seen lying on a pillow, or one that you saw on the street a few minutes ago. The pillow could be one or more of several that you own, maybe in the form of a dog bed on the floor or it might be one that you have recently seen in an ad or website that sells products for dogs. Maybe your mind zeros in on one of the dogs and the same for the pillow.

However, what about the relative location? The preposition "on" means one (or more) Entities are touching, or very close, and one is at a higher elevation than the other. Based on the rules of grammar, and considering the order of words in the sentence, the dog is at the higher elevation. If you had imagined a dog bed, it is very likely on the floor, and the dog is laying on the pillow, and you get a pleasant feeling thinking about the scene, especially if it is your own dog that you are thinking about. Of course, you might have envisioned that the pillow is on the bed or sofa and that your dog is not supposed to be on either of those pillows (or the bed or sofa), and your next mental image might be the less pleasant recalled memory of the last time that you reprimanded your dog for the same exact thing.

In either case, your vision would likely be of a dog above a pillow with the full weight of its body contacting the pillow. But what if, after you had formed your image, the sentence was corrected to say, "The dog is near the pillow." Your imagination would likely move the dog in your mental image to a position close to the pillow. Would the dog still be laying down or might it be standing? Would the dog be a few inches or a few feet away from the pillow? In this case the preposition is less specific, and each person would probably have a different image of the exact location relative to the pillow and the posture of the dog. If possible, you might want to ask some questions to get more details.

Now consider the sentence, "Tina plans to meet Julie before 11:00 am." Reading or hearing this sentence, you don't know where the meeting will occur, but that at some time in the future, Tina and Julie will be in close proximity to each other. If you know these people, you will likely imagine Tina and Julie approaching and then greeting each other. This Event would have a time Attribute associated with it, but how exact would be the time of the Event? It is a bit difficult to tell based on the context of the sentence. You know that it is not at 11:00 am; that would be too late. About how long before would depend on the reason for the meeting. For a normal meeting, it might be just a few minutes. If they are going on a business trip and are meeting at the airport, they might be meeting 30-60 minutes before.

So, your mind would likely create an Event and associated images or a video of Tina and Julie meeting, with a time Attribute that has a time range from seconds to many minutes before 11:00 am. With some questions, or additional information, the time Attribute (and probably other details) for the Event could be revised to be more precise.

Similar discussions could be presented with other prepositions. In general, the prepositions themselves provide an approximate idea about direction, time, location, and space, with varying degrees of precision, as shown with the examples for "on' and "near." Our imaginations can create mental images for sentences with prepositions that are based on our own experiences and the number of details provided in the sentences. As more information is obtained, the memories that we have created based on the initial sentence(s) can be updated to increase the precision of the memory.

The Four Types of Linguistic Communication

There are four main types of linguistic communication between the Self and another person, or other Entities, which could be in the form of written documents or verbal communication:

- Statement-To-Self.
- Statement-From-Self.
- Question-To-Self.
- Question-From-Self.

Statement-To-Self represents incoming information, in the form of written or verbal words and sentences, such as that described earlier. As the information arrives, it must be understood in order to have meaning and add to a person's knowledge. Existing knowledge, logic, and reason are often needed to determine which is the correct meaning of each of the words or word combinations (could be an idiom or jargon). Then the mental images and videos that most closely match the information must be summoned from memory and modified as dictated by the adjectives, adverbs, and other parts of speech in the Statement-To-Self. One or more Events can then be mentally created representing the statement, which would then be stored in memory. If a new Entity, or modified previously known Entity, is described, a new Entity mental structure would be created or an old one modified. Often, the complete information for the Event or Entity is not included in the sentence, so an incomplete Event or Entity mental structure is created. As more information is obtained, more details can be added.

Once the mental images and videos have been created and are being stored, this process is likely to be similar to the person visually observing a similar scene and then remembering it. For example, if you had been to Logan Pass in Glacier National Park and you saw a mountain goat cross the road, your brain would store this information as an Event. If you later read about someone's experience at Logan Pass, in which the author described a group of mountain goats by the side of the road, your mind would likely conjure up the background scene of Logan Pass, modify it by replacing the single goat crossing the road with a group of mountain goats next to the road, and store this revised image/video in probably much the same way that you previously stored the live view of the single goat crossing the road.

The mental processes involved with a Statement-From-Self are, in some ways, the reverse of the Statement-To-Self, but there are significant differences. I will use the earlier example of the horse running through the meadow with mountains in the background. Let's assume that we have this Event stored in memory and we have been asked to describe it in writing. The Event mental structure would encompass the main Entity (horse), the main Action (run), Action Attributes (speed, vigor of the horse, duration), location (meadow, mountains), and time (approximate time of day). The text identifiers for each of these concepts would be determined from each of these parts of the Event. For example, horse, run, rapidly, meadow, and mountains. Keep in mind that each of these Event parts might have more than one word that represents it. For example, it could be horse or Appaloosa horse or stallion. Or run, race, dart, gallop. Deciding which words to use will likely require the use of existing knowledge, logic, and reasoning based on the context of the request for this information, and the perspective of Self and the person receiving the information.

Using the rules of English grammar, the words would be assembled into a sentence (it could be more than one sentence, but I will assume it is one sentence for simplicity). Notice that we need to use our mental capabilities to determine which words are most appropriate, but we do not need to try to understand the Event. Unlike when the statement is To-Self, in which we have to interpret the parts of the sentence to determine the meaning, when the statement is From-Self, the meaning is already known. The effort in this situation is to convert the Event concepts into the most appropriate words joined together by proper grammar. The correct noun must be selected for the subject Entity, as discussed above. Not only does the verb root need to be selected, but also the tense and person of the verb. Adjectives and adverbs will likely need to be included to modify the subject and verb based on the level of detail needed. And pronouns and auxiliary words may be desirable or required, such as articles, prepositions, and conjunctions.

A Question-To-Self is similar, in some ways, to a Statement-To-Self in that, in both cases, an effort must be made to understand what is being said or was written. In addition, a Question-To-Self could also be related to a Statement-From-Self, in that, in many cases, the question is an attempt to obtain information that the Self knows, whereas the statement is the Self attempting to express what it knows, and often a Statement-From-Self is a response to a Question-To-Self.

For a Question-To-Self, the first step is to understand what is being said or was written, very much the same as for a Statement-To-Self (as stated above). However, for the next step, rather than adding new information to the memory of the Self, it must determine if it has in memory the information being requested, or if it needs to first learn something, add this to its memory, then retrieve this information for a response. In either situation, at the point of retrieving information from memory, there would eventually be a Statement-From-Self.

For example, assume that the Question-To-Self is, "What is the color of John's hair?" The Self would search its memory for the Entity "John." If found, the Self would then search John's SubEntities to find "Hair." Let's assume that the Attribute for "Color" is present and it is black. The Self should understand that it should respond back with a Statement-From-Self, which would proceed as stated earlier.

Alternatively, let's assume that the color of John's hair is not known. The next thing to do is to learn this information. At this point, a decision needs to be made on the best way to do this, which is the topic of the latter chapter, "Predictions, Planning, Execution, and Evaluation." A mini mental brain-storming session would need to occur to create a list of the possible activities that could accomplish the task of learning the color of John's hair. This list could include the following potential activities:

- Meet with John (so that the Self can observe his hair color)
- Contact John and ask him
- Ask someone that knows John
- Find a photo of John

Once the list has been created, an evaluation of each activity will be made to determine the best option. The appropriate activity would then be implemented, and if successful, the color of John's hair would then be added to the known Attributes for the SubEntity "Hair" for John. At this point, a Statement-From-Self could be made as a response to the original Question-To-Self, as discussed above.

Finally, the purpose of a Question-From-Self would be to obtain information that is either missing from an existing memory, as in the previous example about John's hair color, or to learn about something completely new. More specifically, some of the types of information that might be desired are:

- A missing Attribute for an Entity or SubEntity
- Missing Group Attribute for an Entity in a Group
- A missing Attribute for an Event
- A previously unknown Entity, Group, or Event
- A rule for logic and reasoning

In each of these cases, creating a Question-From-Self to obtain the information would be just one of the available options, as discussed above.

Emotions

As was discussed in the previous chapter on Needs and Emotions, emotions are often associated with Entities and Events. However, they are also associated with words and phrases. Let's consider the following Statements-To-Self:

- It was a warm sunny day, with the sweet aroma of blooming flowers wafting through the air.
- It was a bit chilly, and light rain was forecast for later in the day.
- The raging storm that evening had already knocked down several trees, the power had been out for an hour, and the backup generator would not start up.

Each of these sentences likely invokes different emotions, although there would be some variation from person to person (unlike your normal individual, a storm-chaser might be ecstatic about the third sentence). As these statements are converted to Events and stored in memory, these emotions would be associated with the Events.

But even though words and phrases are tied to Entities and Events, the linguistic expressions have emotions associated with them that are distinct from the mental constructs of Entities and Events. To demonstrate this concept, consider the next example.

Assume that your colleague has sent you results from an experiment, but his conclusion about the results does not seem to be correct. The mental constructs of the data would consist of Entities and Events that would represent the results that were sent. The Certainty Attribute would be set at a low level for these Entities and Events, indicating that the veracity of the results is not at all certain. You would like to determine whether these results are correct or that a corrected set of results should be sent, by replying with a Question-From-Self or Statement-From-Self indicating that you question the results.

There would be several ways to construct the response, each of which would likely result in a different emotional response by the colleague. Some examples of the response might be:

- "The results are not what I expected."
- "Are you certain about these results?"
- "Your results appear to be wrong."
- "I believe that your results are completely wrong."
- "Your results show that you don't know what you're doing."

You can anticipate how the colleague will feel about each response by pretending that you received these responses from someone else. The colleague's emotional response to each of these statements/questions will probably be similar to what your own emotional response would be. If you tend to like your colleague, you will probably send a statement similar to the first one. On the other hand, if you have an on-going feud, then you might choose one of the latter statements.

When communicating verbally or in writing, with Statements-From-Self, we routinely determine the wording for the concepts that we want to convey based not only on the specific information that we wish to convey, but also by the emotions that we want to invoke in the recipient. Think back to the last communication that you had with a colleague, friend, or relative, and try to remember how much effort was given to your concern about how the recipient would feel about the information that you were transmitting.

In this example, when the colleague reads the response and converts the text into Entities and Events, certain emotions will be created in the colleague's mind based on the meaning. However, when you are thinking about which response to send, your understanding of how the colleague will perceive the response causes your brain to associate emotions with the words and phrases in the sentence based on the emotions that you know will be created in the colleague when the response is read by him/her. So, there are emotions associated with the actual words selected as well as emotions associated with the actual Entities and Events that the words describe.

AI's Understanding of Linguistics

Most types of AI that would interact with humans either by text or verbally must be able to interpret statements and questions that they receive in much the same way as a human would. An AI would need to be able to understand Entities and Events as well as a human does so that they can relate to each other, think, and communicate in a cooperative manner. This would include uncertainty and probability, for example. The AI would, in most situations, need to be able to understand the level of uncertainty in the statement/question, and incorporate this into the resulting Entities and Events added to its memory.

However, the level of complexity required in communication might vary significantly depending on whether the AI will be moving boxes, acting as a caretaker, or finding a cure for cancer. So, the required level of understanding for the AI could vary a lot in much the same way as it does for humans.

Regarding statements or questions made by an AI, they should certainly be logical and either provide useful information to human associates or help to increase the understanding of the AI by it asking questions. If the AI is employed in a technical field, the phrasing of the statement or question might not matter quite as much as if the AI is employed in a customer relations position. In both cases the statements/questions from the AI should consider the emotions that might be elicited by the human recipient, but for the first situation, the concern is potentially irritating a colleague, whereas the second situation could result in a lost customer.

For an AI that has AGI, in most cases the AI should have the same or better capabilities as a human, including those related to emotions, for converting language to and from Entity and Event mental constructs. However, an AI with AGI should be able to more easily handle sentences, and entire documents that have a high level of complexity, than a human, and be able to communicate at least as well as a human.

10

Memory, Rules, Memory Agents, and Rule Agents

"Nothing fixes a thing so intensely in the memory as the wish to forget it."

- Michel de Montaigne

In the previous chapters, I discussed Entities and Events, their details and organization, and their relationships with other Entities and Events. The subject of this chapter is how information that is received becomes converted into retrievable memories, and how patterns are discovered and converted into rules for humans' minds. And what aspects of human memory should be replicated or should be different for an AI.

For the human brain, the process of memory creation is only partially understood. And there are several aspects [34, 35, 36, 37, 38], which include:

- Physiological – the weighting of existing synapses and the creation of new ones.
- Types of Memory – such as short-term and long-term.
- Brain Anatomy – the structures involved with memory formation.
- Organization – how is long-term memory organized.

The physiological aspect involves the strengthening and weaking of synapses, as well as the creation of new synapses and removal of old ones [34]. However, there is also evidence that new neurons are created in some parts of the brain. Also, the neuronal connections tend to be activated and strengthened, or deactivated and weakened, in groups or ensembles. And, as time passes, memories will fade, strengthen, or change as new experiences occur and the synaptic connections change. This activity is referred to as synaptic plasticity.

There are several types of memory [34, 39]. Long-term memory includes explicit, or conscious, memory which can be further divided into semantic and episodic memory. Semantic memory involves general knowledge, such as word meanings, concepts, facts, and ideas. Knowledge of Entities, Attributes, and Groups, as discussed earlier would be stored in semantic memory. Episodic memory involves experiences and events, so it would refer to Events and MultEvents that have been covered previously. Another type of long-term memory, known as implicit, or unconscious, memory is involved with motor skills. And short-term memory allows someone to remember a small amount of information for just long enough to dial a phone number, write it down, or remember it in long-term memory.

The three main areas of the brain involved with the creation of explicit memory are the hippocampus, the neo-cortex, and the amygdala [34]. The hippocampus is responsible for creating episodic memories, although it is not the location where this type of long-term memory is stored. The neocortex is involved with sensory perception, spatial reasoning, and language. However, this is also where semantic and episodic memories are stored. The amygdala associates emotions with memories and helps to form memories involving fear. An area of the neo-cortex known as the prefrontal cortex is responsible for short-term memory. Implicit memory involves the basal ganglia and cerebellum.

Here are four of the main theories of how long-term memories are organized [36, 38, 39, 40, 41]:

- Hierarchies
- Semantic networks
- Schemas
- Connectionist networks

Hierarchies

The theory of hierarchies proposes that information is stored as hierarchical structures, in which there are main categories with multiple layers of subcategories. For example, an organization may have a CEO as the head, with a president, vice-presidents, directors, and group leaders as management positions in hierarchical layers within an organization (see Figure 15). The governmental structure of the United States is divided based on the hierarchical system of federal, state, county, and city governments. This organizational concept can be related to Entities and Groups.

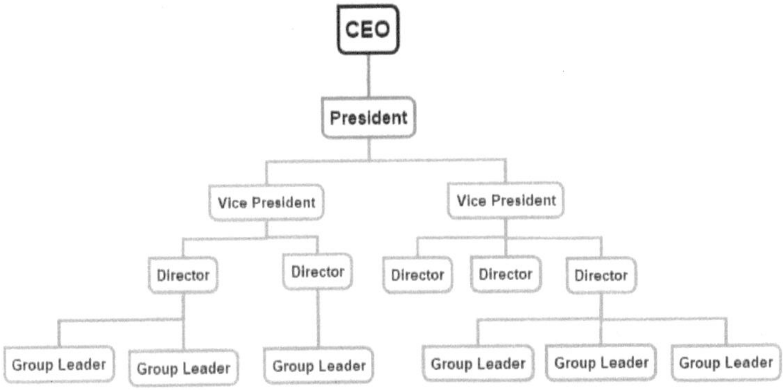

Figure 15. Organization chart showing hierarchical structure.

Another analogy which relates to Events and MultEvents is a book, which usually has the following hierarchical layers:

- Chapters
- Sections
- Paragraphs
- Sentences

In certain ways, the progress of our lives is hierarchical. At a high level, a typical life might have the following sequential stages or MultEvents, which could be considered as chapters:

- Spend most of our time at home for the first few years of our life
- Pre-School
- Kindergarten
- Elementary School
- Middle School
- High School
- College
- First Job
- Second Job
- Third Job
- Retirement

Each of these "chapters" would have "sections," such as first through fifth grade for elementary school. And "paragraphs" might be what happens on each day. Our lives are very complex, so they could actually be divided up into numerous different hierarchies, some of which might run in parallel. For example, the purchase of a house while in one of your jobs could be divided into hierarchical steps. And the same thing would apply if you had children, or even a pet.

Other concepts that are hierarchical are:

- Taxonomy
- Body parts
- Machine parts
- Computer folders
- Library classification
- Maslow's Hierarchy of Needs
- Family structure
- Language family tree
- Color wheel

Notice that some hierarchical structures are conceptual, some are physical, and some may be both. Taxonomy and Library Classification are conceptual, and Body and Machine Parts are physical. Government structure and organization are both because the organization of the various jobs is conceptual, but the actual division of the geography into states, counties, and cities is physical.

I have previously discussed the concepts of specific vs generic examples of Entities, which is essentially a hierarchical concept. For example, let's say that Pete's dog is a black labrador. The first time you see it, you will notice certain things about its appearance, temperament, activity level. When you see it again and again, you will likely observe other aspects of these attributes. The mental images and videos are stored in memory as representations of Pete's dog. So, to really know Pete's dog, you have to recall many of these memories

Now, let's say that Don also has a black lab, which you encounter frequently. Your mind will store up mental images and videos of Don's lab, so you will have many mental examples that combine to create "knowledge" of Don's dog. However, Don's and Pete's dogs will have slightly different physiques and personalities. If you observe several more black labradors, you will notice that they are all somewhat different, but likely still have many things in common, such as hair color and type, general body shape, obedient and playful personalities. To fully understand the black labrador breed, you need to remember the various traits of each individual dog and then the combination of all of the black labs that you have been exposed to. So, the first hierarchical level is a given experience with one dog, the second hierarchical level would be all of the experiences with a given dog, and the third level would be the combined impressions of all of the black labs that you know or have seen.

Schematic Networks

These are somewhat like hierarchies but are more random. They represent concepts that are related, but in a less structured way, and the concepts are not necessarily of the same type. For example: tree, leaves, autumn, raking, garden tools, gloves, hand, fingers (see Figure 16). A stream of consciousness could be represented by a similar structure. Hyperlinks on webpages are another example of semantic networks.

Figure 16. Example of Schematic Network.

Schematic networks represent large parts of the overall structure of our memories. Think of an Entity, such as a specific person named Alice. She lives in Seattle and works at a company there. Jacob also works at that same company, but is from Chicago, where he grew up. In high school, he was in track, so he ran a lot, and he had a friend that he would run with in the evenings, who had a dog that ran with them. The dog also liked to catch tennis balls. But the dog would tend to chew up and destroy the balls, so his master had to go to the pet store a lot to buy replacements.

We probably all do this stream-of-conscious type of thought frequently, and it is likely a key aspect to creativity, since it allows many disparate thoughts to be considered when someone is trying to find a solution to a problem.

Schemas

These are somewhat like schematic networks, except that they show the relationship between Entities or concepts based around a specific topic. For example, parts of a human body, or parts of an engine (see Figure 17). Often schemas are somewhat like a hierarchy with only two levels, the main concept and the sub-concepts related to it. However, they can have several or many levels. In the example below, "Eyes" could have had shown all of the parts of an eye (cornea, retina, lens, etc.), and retina could have had all of its parts (cones, rods, blood vessels, etc.).

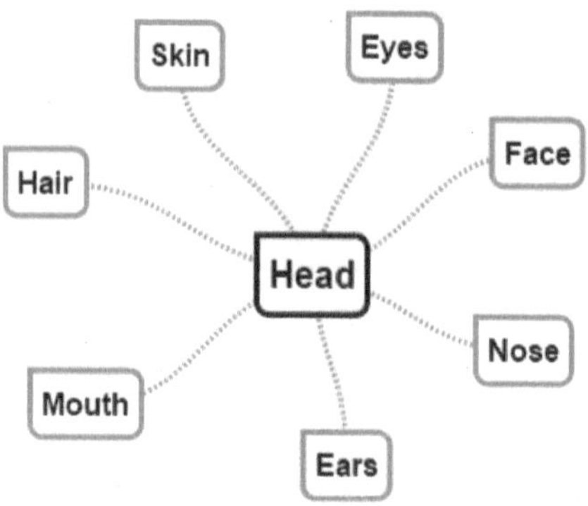

Figure 17. Example of Schema.

The organization of information related to Entities can be considered as physical and conceptual schemas. The diagram above is an example of a physical schema and could represent the parts of most Entities. Associated with the Entities are the various types of Attributes, including current Location, possible Actions, previous Events, associated Groups, all which can be organized and represented as one or more schemas.

We can imagine combining these concepts of memory organization, as shown below (see Figure 18).

Figure 18. Example of combined hierarchy, schematic network, and schema.

Connectionist Networks

The fourth theory on memory organization is that of connectionist networks (see Figure 19). This concept resembles the models for artificial neural networks, in which there are layers of nodes for input, output, and intermediate, or "hidden," nodes. In this case, the nodes represent concepts, instead of neurons. Each of the nodes for one layer are connected to each of the nodes in the adjacent layer. As learning occurs, the weighting changes for each of the node-to-node connections.

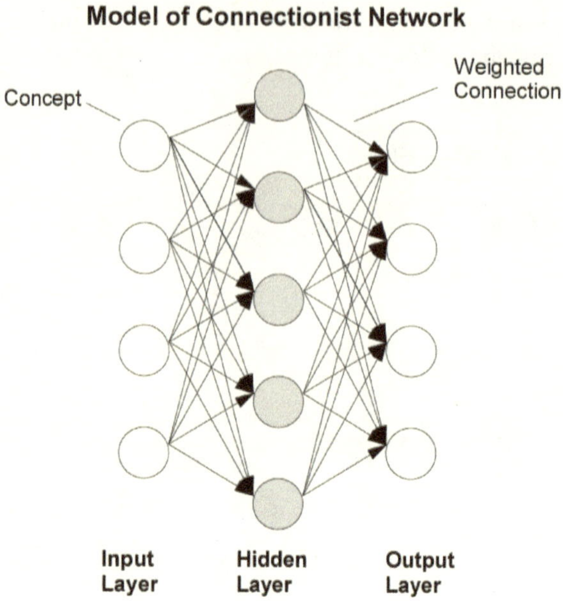

Figure 19. Connectionist network concept.

Other Types of Long-Term Memory

In addition to those just discussed, I believe that there are several other basic types of memory organization, some of which are listed below:

- Sequence
- Causation
- Associations
- Pattern
- Motion

<u>Sequence</u>

Much of what we learn and remember is a sequence that is in a specific order. This includes the alphabet, integer numbers, recipes, driving directions, assembly instructions, operating procedures, the hours for a day, a series of Events, etc. To appreciate the significance of order in relationship to how your brain organizes and remember sequential information, try to say the alphabet backwards. Often, the relationship between the elements is conceptual, but can have physical or mathematical associations, such as for mile-marker signs or the alphabet as represented in a textbook, which are separated by a specific physical distance, or bits on an Analog-to-Digital chip, in which each progressive bit represents a voltage 2 times higher than the previous one. An example of a conceptual sequence would be a recipe. The steps much be followed in a specific order, but there is usually no physical relationship between the steps and the time between the steps may or may not be important.

Causation

This is a type a sequence in that two (or more) Events occur in order. However, it is different in that with causation, the first Event causes the second Event to occur, whereas with the alphabet, 'A' doesn't cause 'B' to be present. Or for a food recipe, a given step in the recipe probably causes some physical activity to occur, which is itself a cause-and-effect relationship, but the given step usually doesn't cause the next step in the recipe to occur. So, causation requires a specific type of memory organization in which if the first Event occurs, the second Event is very likely to occur after the first Event and can only occur in that order. As we all know, your grass being wet will not cause it to rain.

Associations

This type of memory organization is used to store the relationship between Entities and may or may not have a pattern associated with it. Examples are the Entities in a room (furniture, carpet, bed, lights), people in a store, and trees in a forest. To remember the contents of a given room, your brain must be able to not only recall the specific Entities, but also their locations relative to each other, even though these associations might be quite random.

Patterns

These represent associations of Entities that have some relationship between them which is predictable, such as equal distances, mathematical relationships, or time sequences. For this type of memory organization, your brain needs to be able to remember what the pattern is and how this relates to the relationship between the Entities.

Motion

There are actually several types of motion that need to be stored in memory in relation to Entities and Groups. One type in involves the movement of one or more Entities relative to other Entities or a location, such as a flock of birds flying or car moving on a highway. Another type of motion is the coordinated movement of several parts, such as in an engine.

Memory Organization

Throughout this book, I have discussed conceptually how information is likely organized in the brain with regard to Entities, Groups, Events, linguistics, and logic and reasoning, including the ideas presented in the chapter. But now I would like to discuss how information is brought into our brains or an AI and is organized and stored in a way that it can be retrieved and be useful in the future.

We want to use the human brain as a model, since it is the most advanced thinking machine that exists. This doesn't mean that it is the best information processing machine that is possible. But it does work, so it represents a good starting point. However, there are at least two main reasons that we will not be copying exactly how it works. The most obvious one is that we don't understand the brain well enough to do this. But we do understand many, and maybe the most important, conceptual activities that it is able to carry out. For example, we know that the human brain can store the details of an Event, is able to recall the details in order to communicate them to another person, and it can also use them to create new logical relationships and for planning, even if we do not know exactly how it does these things. Our goal is to develop an AI with AIG capabilities, so if we are able to create an AI that can mimic the most important thinking tasks that a human brain is capable of, but uses software code instead of neurons, we should be able to accomplish our goal.

The other reason that we will not be exactly copying the human brain is that we probably do not want to if we are able to improve on certain aspects. For example, an AI with AGI capabilities needs to be able to observe the world and remember what it sees. The human brain can only store a limited amount of detail from its visual observations, although some people are better at this than others. And the human brain can only store and retrieve information at some limited speed that is much slower than is possible electronically. An AI should be able to remember what it sees in much greater detail than a human brain, and it should be able to do it much more quickly.

I believe that the analogy between the bird and the airplane is very applicable here. The goal for both is to fly. And there are certain concepts from the bird that must be replicated in the plane, such as the use of wings with a minimum amount of surface area and type of curvature that will allow sufficient lift to be created as air flows over the surface. However, whereas the bird flaps its wings to create propulsion, the airplane uses rotating propellers or turbines (as for jets), and whereas the bird twists its wings to allow rotation along the longitudinal axis, the airplane uses ailerons.

There are many types of memory organization that occur in a human brain, some of which are understood to some level of detail and other types that are poorly or not understood. But, as I mentioned, our goal is not to recreate the workings of a human brain, but to create an AI with AGI capabilities that can process information in a manner similar enough to that of a human brain that the AI can essentially carry out the same tasks as well or better.

For the remainder of this chapter, I will propose memory organization structures and activities that should be able to accomplish this. Many of these ideas are based on what we know, from our own experiences, that a human brain does. Other aspects are based on speculations of what the human brain must be doing. And yet other conceptions might be considered as speculative enhancements to what a human brain does.

At the beginning of the chapter, two main types of long-term conscious memory were discussed, which are semantic and episodic. However, as was discussed in previous chapters, there are also types of memory that store the relationships for patterns, causation, analogy, math, grammar and Groups. In *Words and Rules* [42], Steven Pinker discusses aspects of how the human brain handles language. He argues that word roots and their meanings are stored in a type of semantic memory. However, rules are applied to determine when a -d or -s, for example should be added to a word root for past tense or plural words. These rules can be applied even to words of which we don't know the meaning. For example, if I told you that I have one "plubb" on the table, and asked, "What would you call two of them?" you would probably say, "Two plubbs."

In addition to semantic and episodic memory, I believe that there is also associative and interactive memory. We can think of semantic memory as storing information on Entities and episodic memory dealing with Events. Associative and interactive memories would then pertain to Groups and MultEvents. In other words, the relationships between Entities and between Events. Rules are derived from these relationships.

Let's look at an example of how a grammatical rule might be created. If I see one shoe and it is called "shoe," but that two or more shoes is called "shoes," my brain can create an associative memory for the singular and plural version of "shoe." Then I see the same thing with sock and bicycle and plate and dog. These are specific associative memories. But my brain can generalize these memories to create a rule that says if there are more than one of an Entity, add an "s" to the end of it. Of course, the rule is a little more complicated that this because sometimes an "es" is added (kiss/kisses) and there are irregular words such as "datum/data," "goose/geese," and "ox/oxen." The "es" would involve a combination of a memory and a rule, but most of the irregular words would only involve memory.

In the next sections I will first discuss types of memory, followed by a discussion on rules.

Background Information

As we go through the day, information is constantly arriving at our brains from our five senses. This is like a continuous movie but instead of just the sight and sound being synchronized, touch, smell, and taste are too. However, this information is initially being stored in a type of short-term memory. Some of what we perceive will be transferred to long-term memory, especially if we focus on certain aspects of it, or the perception was accompanied by a frightful event (such as almost falling off a cliff or just remembering that college final exam that you forgot to attend).

Before theorizing about how the human brain organizes information as it is received, let's consider some common awarenesses we all have about everyday life. Imagine walking into a large room for the first time with a lot of furniture, art, photos, maybe bookshelves or a cabinet with mementos. Or a new store with seemingly limitless shelves of products. Or a museum with its many artifacts. Your brain will likely be a little overloaded with all of the details and you will only be able to focus on a few things. After leaving, you will probably not remember most of the details, even immediately afterward. There will be a few things that you might recall, especially those things that are similar to something that you are already familiar with. However, you may not remember where in the room they were located.

We don't have a video recorder in our heads that stores everything and later allows us to re-observe the scene. This is because the visual input goes into short-term memory. Whatever our minds can transfer to long-term memory is saved, but the rest is lost, just like the unfamiliar phone number that you looked up earlier today. This is why the digital cameras on our phones are so amazing. They allow us to create that detailed long-term memory that our brains are not capable of doing.

If we later revisit the same room, we will likely notice more details, more objects, which will be added to our established, but sparsely populated, memory of the room and the level of detail will be enhanced. With repeated visits, we may be able to remember in detail the main items in the room, their location, and if something has been moved, is missing, or is new. And our brains will likely also timestamp the visit, recording when you were there and for about how long.

The first time that you entered this room, your mind established a memory of it, even though it only contained a few details. With each return, it was enhanced with additional details. If the room happened to be in a large museum with many rooms, you would probably need the map provided with your entrance tickets to navigate the museum, at least on your first visit. For the next few visits to this museum, you will probably still need the map. Even though your mind is able to remember more details about a given room, you likely do not remember the location of that room relative to the others or the museum entrance. Of course, some people are faster at learning directions and special locations, but with every new location that we are exposed to, we all must create a mental map that represents how specific locations are related to surrounding locations and the proximate world.

Dogs provide a window into these same concepts. My three dogs love exploring our wooded property, which has many trails throughout, which were originally created by deer, elk, and some moose (that, by the way, still use them). Even with all of the details around the house, such as the driveway, flower pots, gardens, and the surrounding forest, if there is a package left at the house or a branch on the driveway, my dogs will go up and sniff it, indicating that they realize that the new item is not part of the normal scene.

When we go for walks on the trails in the woods, they do not wear leashes and are free to wander ahead or behind me. They have learned shortcuts using the network of trails. If I lose track of one of the dogs, she will find us and catch up in a few minutes. Sometimes, I will make it back to the house before she returns to us, but each of them will make it back to the house in a few minutes. This means that my dogs have a mental map of the property and know where they are relative to the house, myself, and the many trails.

Regarding the room examples, only a small amount of the total information arriving to a human brain is processed. However, an AI could potentially learn about a room much faster than a human can. With enough memory, processing speed, and adequate sensors, an AI should be able to remember everything that it sees, hears, smells on its first visit, in much less time than most humans are able to, no matter how many times they have been in the room. And, in addition, the AI may have sensors that a human does not, such as for infrared, ultraviolet, subsonic, ultrasonic, very low or very high temperature, or even radioactivity. These extra-human capabilities would provide the AI with a richer understanding of the room and its environment than is possible for a human not equipped with these instruments.

Memory Overview

As information arrives in our brains, we compare the new knowledge with what we already know in order to identify Entities and Actions that we are observing. We are also identifying sounds, smells, touch sensations, and possibly tastes. But we are also predicting movements. For example, if we are outside and see a flock of birds flying overhead, we may try to predict where they will disappear beyond the tops of the trees. Or, if someone throws a ball toward you, your mind will try to predict where the ball will be so that your hand can catch it (and tennis, baseball, basketball, football, and soccer players demonstrate just how good the brain can become at motion prediction, especially when the pay is substantial). Or, if a car just ran a red light and you are maneuvering your car to avoid being hit.

You might also be trying to assemble something, so your imagination is rotating the bolt you are holding in order predict how you should move your hand in order to line it up with the nut that is welded to the bicycle frame in front of you.

Your mind also has a concept of where you are relative to your surroundings so that you know how to approach Entities that you want to and avoid other ones that might be dangerous. If, for example, you have a sharp piece of metal on your workbench for a project that you are working on, you want to be careful as to how you approach it. Or if you are hiking in the mountains and pass by a steep drop-off, you will make sure that your body does not get too close. Or with the car running the red-light example just discussed, you make every effort to avoid an accident.

We may be trying to carry out a plan, such as going to the grocery store. In this case, we are comparing the immediate information that is arriving at our brains with the plan that we made earlier and continuously making judgements and adjustments to our Actions.

Or we may be reading a non-fiction book, in which our minds are retrieving information in order to understand what we are reading, and then adding, often with the assistance of our imaginations, new knowledge to our existing mental databases.

So, as we go through the day, we experience the world around us, use current knowledge to understand it, but then add this new information to our memories as enhancements to Entities and Actions that we are already familiar with. But these details, as mental videos and images, become a sort of chronology of what we did during the day with both time and location stamps. And we use this incoming data to help keep ourselves safe and to assist us with carrying out plans we have previously made.

I would like to now go more in depth into these concepts and theorize about how the human brain organizes this information. But we need to keep in mind that the ultimate goal is to create an AI with AGI abilities, which does not necessarily need to function in exactly the same way as a human brain. So, a theory about how a human brain organizes information might provide a means for how an AI can do it, even if the theory is not exactly a blueprint for how the human brain works.

Real-Time Personal Environs Memory

Let's first review the earlier discussion in the chapter on Entities, Actions, and Events about how information is organized and stored for an Entity. It will normally have several layers of SubEntities, such as body, head, eye, retina, etc. And for an Entity, and its SubEntities, there will be many types of Attributes associated with the Entity and its SubEntities.

When we think of a person that we know well, we can visualize many images and videos of them from many angles, wearing different clothes, doing different things, at different times. And most of us will visualize the person in 3D, including the videos. Each time that we saw this person, more images and videos were added to their mental database. So, there is a mental construct for this person that has links to all of their memories across time. And we are not just storing information about the person as an Entity, with all of their associated Attributes, but also all of the Actions of this person. Walking, running, sitting, eating, talking, etc.

And each of these memories is time and location stamped. We can remember approximately, or sometimes exactly, each time we saw the person, and probably where we were when we saw them at that time. And over short periods of time, the person's basic appearance would normally be fairly consistent. But each day, they will be wearing different clothes, and over a given medium time period, they might grow or cut off a beard, cut their long hair short, change their hair color, gain or lose a lot of weight, etc. If we know the person for years or decades, we will also see them age, and we will have memories of how they looked at different times in their life.

And the same goes for the Actions. The type of activities this person engages in will change with time. They may become fitter, or injure themselves and walk with a limp, or they may play an instrument, which they become better at with time.

When we think of this person, our minds don't create one 3D image that is an average of every time we have seen them. Nor do we scan through every image that we remember about this person. So then, what kind of visual image, or images, of this person are we seeing in our minds?

Let's consider how we think about dogs. Most of us have probably observed quite a few different dogs breeds. And if we have been fortunate enough to have seen several dogs of the same breed, maybe a common breed such as a black labrador or golden retriever, for example, we have likely noticed that each one looks a little different, regarding its size, appearance of its hair, possibly some white on its toes or chest (black lab), the way it walks, and maybe its bark. And the personality for each dog will be somewhat different, although there will be many similarities for a given breed.

When we hear the word "dog," does our mind conjure up a dog that is the mathematical or physical average of every dog of every breed that we have ever seen? Not likely. You will probably think of your own dog, or a friend's dog, or one that you saw earlier today. And if you hear the word "black labrador," but you do not have one or have not seen one recently, your mind will likely remember the one you saw a month ago, or on TV, or in a magazine. You may even think of the last two or three different black labs that you saw. You will likely not remember every detail of any of these. And I don't believe that your mind is creating an average of the black labs that you have seen. Instead, you are probably thinking of specific dogs. And although you probably do not remember every detail of them, I do think that it is possible that you might remember the white tips on the tail of one dog and the personality of another, and blend these together. But this would be a kind of mixing, rather than averaging.

Likewise with the human friend. When you think about them, the first mental image or video that appears will probably be of the last time you saw them, maybe when you went out to eat lunch, or when you were together at a party or special event. However, you can probably think back to what they looked like a few years ago, or before they cut off their beard, or when you were in college together. You may not remember every detail, but you will likely not average images together.

I think that remembering scenes, or Events, is similar. We talked earlier about experiencing a new room or location, and how more of the details would be added to our memory of that space with each new exposure. But let's think about a room that you might be very familiar with for several years, such as your garage. Assume that when you bought your house 10 years ago, the garage was empty and did not have any workbenches or shelves. But through the years, you added cabinets, shelves, and workbenches. And, of course, you accumulated a lot of nice tools because each time that you made improvements to your garage, or used some of your tools to make or fix something, you were able to justify the purchase of that next shiny new tool (of course, you'll be able to accomplish so much more with that new band saw).

If someone asked you about your garage, an image of how it currently looks would probably pop into your imagination. But if you were to explain to them everything that you had done to it, mental images at various points in time would appear, earlier ones showing a bare garage, some slightly later with the new workbench, then another with the cabinets, and so on. Likely, the details of the earlier mental images would not be as good as the later ones, since you would have forgotten some things.

An important point here is that for the friend, the dog, and the garage, these Entities each have their own mental construct, but there are many images and videos that are associated with each of the Entities which were acquired though day-to-day observations of each of them. But, as we go through the day, we are not just seeing these Entities by themselves. For the friend and the dog, they are just one part of a scene that probably includes trees, grass, sky, sidewalks, cars, etc. Even with the garage, which is a location in itself containing many Entities, it is part of the house and has an outside area in front, so the garage is often part of a larger scene. Therefore, our minds need to notice and identify each of the Entities in the scene and associate its image with the corresponding mental constructs that already exist for those Entities.

Another important point is that our minds create and store mental images in the same dimension as the Entity, which was discussed in the chapter on Entities. When we see a person or dog or car or building or mountain, we can actually only see it in 2D, but if we can see the Entity from several angles, our mind is able to convert the 2D images into 3D. And even if we observe a new Entity, our minds can use analogy to guesstimate what the other sides look like and create a 3D mental image.

For example, if you were familiar with horses and then saw a zebra for the first time, but only just a quick glance, you would certainly be able to visualize what the other side of the zebra probably looks like. Or if you had never seen a zebra or a horse, but you were familiar with cows or goats or sheep or dogs, your mind should be able to use analogy to imagine what the other side of the zebra looks like (probably just like the side you are looking at) and use analogy to create a 3D mental image. Even if you only saw the front of the zebra, you could probably imagine what the back end of the zebra looks like.

As another example, maybe your dog is walking around in your living room. If she walks part way behind the sofa so that you can only see her front half, your mind immediately imagines what the rest of your dog looks like. And if she goes completely behind the sofa, you can imagine the entire dog, maybe where she is, especially if there is something there that you might not want her to get into. You are not only imagining the side of your dog that was facing you when you last saw her, but also the other side and the entire dog. You actually have 3D images and videos of your dog, probably many of them, some of which you are retrieving. But when your dog is not behind something, you are also adding to memories and the mental construct of your dog, which can be later referenced along with the many other previous memories.

So, we may also be storing in memory some of the 2D images, but some or most of the images and videos that are stored are as 3D images. With these, we are able to use our imaginations to mentally move and rotate different kinds of Entities and estimate how they will interact with other Entities, such as will one fit inside the other or fit through an opening or fit into certain clothing.

Now, let's think about how our minds create a chronology of our day-to-day activities. As we go through an average day, many of the things we see are familiar to us, such as our house, car, our workspace (office, lab, shop) and even the route to work. But there may be some experiences that are completely new. There may be an accident on the highway, and we must detour to an unfamiliar road, or we may stop at a new restaurant for lunch or a store after work that we have never been to. For the locations that we are familiar with, if there is something that has changed, we will probably notice it, even if it is a small detail. However, for the unfamiliar places, we may not remember much of what we saw, as was discussed earlier.

What we are really experiencing is a series of Events that are strung together as a MultEvent or several MultEvents. Getting up in the morning could be an Event. Taking a shower, getting dressed, eating breakfast, walking to the car, etc. could all be Events. As you observe your surroundings, using all of your senses, during each of these Events, your mind is, almost subconsciously, identifying the main Entities around you. For any of the rooms of your house, you already have many memories of that location, with most of the Entities identified and labeled. But if there is something new, such as a new shirt hanging on the doorknob, your mind will use analogy to identify it at least as a shirt, or possibly as the shirt you bought last weekend at a specific store for a certain price. When that shirt was purchased, a mental construct was created for it as a specific Entity, with several Attributes, and it is now being referenced as part of the Event that you are currently experiencing this morning. And as you store the memory of your bedroom as one of the many Events you will experience this day, the numerous Entities in that room, including the shirt, will be referenced as part of that memory.

You would have many memories of your bedroom that exist in your mind essentially as a Group of numerous Entities, in which the

Entities are related to each other by their relative locations in 3D space. As you think about this room, you can imagine it from many angles and visualize moving through it, so your bedroom exists as a 3D Group and mental construct in your mind. And each of the Entities in the room, the bed, dressers, mirrors, pictures, lamps, are 3D Entities with many Attributes, including their name, size, shape, color, weight, value, etc. Each of these Entities are mental constructs that exist in your memory. When you imagine your bedroom, your mind is retrieving it as a 3D image, which is based on 2D images, since that is the way you see it, and videos of when you have walked through it. So, there is a mental construct of your bedroom that has links to the 3D, 2D, and video images that your eyes have seen, and your mind has created.

And each of the items in your bedroom are Entities that link to the many 2D, 3D, and video images for them. So, as you imagine this room and you are able to identify each of the things in it, each of the mental constructs for the individual Entities are being referenced through links. And when you think about a specific Entity in the bedroom, you are able to reference all of the images that are associated with it.

I have been focused on the visual and 3D aspects of Entities, but each Entity would also have memories as Attributes for the other 4 senses, engineering concepts (heat capacity, viscosity, etc.) value, quality, emotions, etc. where applicable.

In addition to your bedroom being a Group, it is also a location, as is your house, the street in front of your house, a park, a lake, a store, a city, or a state. When you think later about getting ready for work, and you think about where you have been, you will remember that you were in your bedroom, as a location.

What if this morning there was nothing different in your bedroom, would your mind still create an entirely new memory of that room? We know that you will remember what you did in your bedroom this morning, so that, in itself, is a new memory. But, as was discussed earlier, you will remember many more details about your bedroom this morning than you would if you had spent that time in an unfamiliar room, such as that of a relative, or a hotel room. This means that you are referencing existing memories of the room. And those memories of the room consist of memories of the individual Entities related to each other, location-wise, as a Group.

I believe that each time that you walk through your bedroom, a new video and images are recorded in your mind for that room. In addition to the construct of the bedroom as a Group, the room contains many Entities, which each have their own mental construct, which are members of that Group. And the new video and images become reference material for that Group and those Entities.

Allow me to summarize these points, still using the bedroom as a reference. I believe that as you walk into your bedroom, your brain matches the main Entities in the room to existing memories of those Entities, so that you can almost instantly recognize and understand them. For each of the Entities, many of the memories, as 2D and 3D images and videos, are essentially links to previous Events that involved those Entities. These memories are used by your brain to help create the current video which is the new Event. And then the new Event becomes a memory, by way of links, for each of the main Entities in the current video. So, in this massive schematic network, the new Event is referencing existing memories in order to help understand and create the new Event, but then becomes a memory itself and a memory, via links, for the main Entities and the bedroom as a Group.

A bedroom makes a good example for thinking about Entities and their relationship to each other as a Group (the bedroom), but not a lot of motion is involved unless another person or your dog enters the room. But most of the locations that you observe each day have some or a lot of movement (a park with birds and squirrels or a major interstate during rush hour, for example). What our minds are processing and remembering with each of these scenes are Entities, Groups, and Actions in a given location and time, which constitute Events. For Entities, what we are experiencing and remembering are their Attributes. For Groups, we are observing and remembering the physical relationship between the Entities. And for Actions, we see and remember the innumerable types of movements of the Entities.

The Events that occur throughout the day are time and location stamped and then organized so that they are in chronological order. That way, you can recall what you did, when and where you did it, and in what order. And when you think of a given Entity that you saw during the day, you will be able to remember what the Entity was doing, and when and where it was doing it. I also believe that the Events are organized like paragraphs, chapters, and volumes for books. During a given day, we can remember getting up in the morning, getting ready for work, traveling to work, working in the morning, attending meetings, eating lunch, etc, as a series of Events. And then we might group these activities by the five days of a given workweek. At a higher level in the hierarchy, we can group all of the Events together that occurred in the each of the various jobs that we had throughout our lives. Or before we were married and then afterwards.

Everything that we experience first-hand is essentially a sequence of Events that our mind groups in ways that help us to understand the world and our place in it. All that we know about Entities and their Actions are ultimately based on these first-hand sensory experiences. The Attributes of any given Entity are observed each time we see and interact with that Entity. We may read about that Entity, or one that is similar, and learn things about it that we have not personally observed. However, to have been able to understand what we read and relate it to the Entity, we had to have personally experienced that Entity, or a similar one, in order to know enough about it to know how to add additional information about it in our minds. If we read about a fictional character in a novel and learn about many of the imaginary person's traits, we could say that we know a lot about that person, even though we have never seen or talked to this person. However, our ability to create a mental construct of this person is based on having personally experienced many other humans and making analogies and extrapolations from the real humans we have known to the fictional one in the novel. Our ability to imagine a person is based on having seen, heard, touched, talked to and otherwise encountered other humans, through some or all of our five senses. This same point applies to other Entities, to Groups, to Actions, and to Events.

Real-Time Self-Based Memory

As we move around our environment, and observe the world through our five senses, there is another type of experience that we are having and memory that we are forming, which is that of the Self. This involves the thoughts, sensations, and emotions that we experience in response to our surroundings.

For example, if we bump our hands on something or touch a hot surface, we may sense pain. If we see something that we like, it will likely give us a pleasant feeling and make us want to stay longer. But if we smell something extremely putrid, we will probably want to leave the area as soon as possible.

What is described here is a response system that helps to avoid dangerous situations, but also assists to avoid unpleasant situations, or creates memories for desirable Events that you might want to repeat in the future.

Let's assume that you have a wood-working shop and that you like to build furniture for a hobby. As you enter the shop, you notice a cord on the floor that you forgot to pick up after the last project. You realize that you could trip on it and endanger yourself, so you stop to pick it up. As you reach to hang up the coiled cord, your hip presses hard against the sharp corner of the workbench giving you an unpleasant sensation. After measuring and marking the initial boards that you are going to cut, you turn on the table saw. At this point, as you begin to cut each piece, your mind is very focused on not getting any of your body parts, especially fingers, too close to the blade. After assembling the piece of furniture that you are building, you stain and polyurethane it. From previous projects, you remember the unpleasant odor from these chemicals, and you know that they are harmful to smell, so you wear a respirator that absorbs them from the air. When you complete the project, you appreciate your good work, which gives you a pleasant feeling, and you look forward to your next project.

As you proceed through the project, you make observations of the room, where each of the tools is located, the operation of the table saw and other tools, the sawdust streaming from the blade, etc. Your mind's real-time personal environs memory stores these Events in order. Simultaneously, your real-time self-based memory is storing the sensations that personally affect you, as well as the emotions that you are experiencing. And all of these memories are time- and location-stamped.

As another example, image that you are giving a presentation in front of a moderately large audience. It could be for a project at work, or at a conference, or maybe a job interview. You tried to prepare well, and you know that the participants will ask questions when you are finished speaking. Your personal environs memory is recording the facial expressions and demeanor of the audience members, stragglers entering and finding a seat, and how well your slides are displaying. At the same time, your self-based memory is recording that you feel a little nervous, but confident at the same time, the temperature is a little high and bordering on uncomfortable, you think ahead to your next sentence while speaking the current one, you are monitoring your rate of speaking, volume, and tone, and you wonder if the audience members are grasping all, or at least the most important points.

I want to point out that in each of these examples, there is more occurring than just remembering incoming information and emotions. Your brain is also analyzing information, reacting to it, and planning actions. This will be discussed in more detail in the next sections.

Would an AI need to have real-time self-based memory? Certainly, especially if it was in the form of a robot, humanoid or otherwise, but not exactly the same as for a human. The AI would need to be aware of itself in its environment so as not to cause harm or to allow itself to be harmed. However, it would likely not have human-like emotions, so it would not sense fear or nervousness or joy. The overall purpose of emotions in humans is to make them do more of what they feel good about and avoid situations that cause them to feel bad. So, an AI could have pseudo emotions programmed into it that would provide positive feedback if its activities went according to plan, or negative feedback if not. This would allow the AI to self-train to do more of the correct things and less of what it should not be doing. And the AI should store in memory its self-based experiences so that it can later associate where it was and what it was doing with how the AI interacted with its environment.

Linguistic-Imagination Memory

During our average day, we read and communicate frequently. As I discussed in the chapter on linguistics, when we read or listen to someone speak about Entities and their Actions, our imagination converts the words and sentences into images and videos based on ones that are retrieved from our memories and modified according to details of their descriptions. These Events, which are created in our minds to represent the linguistic information we just read or heard, are stored as mental constructs much as if we had experienced them firsthand (although our minds know the difference between something experienced firsthand and something that we read in a book).

These "new" Events might be about someone or something or somewhere that you don't know or have never seen and will be stored as entirely new Events. However, alternatively, this new information may provide additional details about Entities or Groups that you are already aware of. And sometimes these details will be about things that cannot be observed even when directly interacting with the person or other type of Entity. For example, you might read that a well-known person has cancer, or read about the construction of the internal parts of a hydroelectric dam that you have seen from a distance, or read about how an engine works. For each of these situations, the main aspect being learned has not or cannot be directly observed.

Reading allows us to add information that we could not directly observe, including for an Entity that we are already aware of.

If you are reading about some Event that occurred, (e.g., in the newspaper) or reading a novel, your mind will create three sets of timelines. Imagine that it is the weekend, and you have some time to catch up on the novel you have been reading. You get a cup of coffee and something to nibble on, then head to the sofa, with its large pillows, located in front of the roaring fireplace. Your dog follows you and lays on its pillow next to the sofa. After getting comfortable, you flip through pages to find where you left off. You notice that it's about 1:30 pm as you start reading. You periodically take a sip of coffee and eat a bite or two. After a short time, you see that the fireplace needs another piece of firewood, so you get up to put it in. When you come back to the sofa, you change positions and continue reading.

As with other activities throughout an average day, your mind is recording your surroundings, the location, and the time, as part of the Real-Time Personal Environs Memory. And you are simultaneously keeping track of the position of your body on the sofa, the taste of the coffee, and your proximity to the hot surfaces of the fireplace as you place another piece of wood inside, in accordance with the Real-Time Self-Based Memory. But, in addition, as you read the novel, your mind is creating and storing Entities, Events, locations, and times/dates from the story. The novel might be set in the present day, or in the past or future. It will likely not provide exact years or dates, and for each day in the novel you may only have an approximate idea of the time, maybe as "it's morning," or "it's afternoon." However, your mind will record the Entities and Events described in the story as you progress. If the novel proceeds chronologically, then the Events will follow each other in order. There may be several locations, so Events and storylines may be occurring, and recording by your mind, in parallel. And the story would be divided into chapters and paragraphs in your mind, as in the book.

So, your mind is recording the information from the current here and now of your surroundings in your living room and simultaneously creating a world inside your brain that is derived from the storyline of the book. There would not necessarily be any link between the action in your living room and that described, and recorded in your mind, from the book, unless perhaps you spilled coffee on Page 71 just as the main character in the story was making her escape.

What is taking place in your mind as you are sitting on the sofa reading is very complex. On the one hand, you are mentally recording images and videos of your surroundings and organizing them in your mind as you mentally reference previous Events to help understand what you are perceiving. Simultaneously, the words on the page are causing your mind to retrieve references to existing Entities and Events and use analogy and other logic to create an entire world inside your mind. And when you later recall your experience in the living room and the Events in the novel, your mind knows which was real and which was fiction.

An AI should be able to perceive all of the sensory information from the living room in greater detail and faster than a human. And it should be able to visualize the story from the novel in the same way as a human, but more precisely, depending on its own background knowledge. But it also needs to be able to distinguish what is real and what is fiction, just as the human can.

Published Images and Videos Memory

But what is happening inside your mind if you are watching a movie based on the novel instead of reading it? Let's assume that you're watching a movie that is a very true adaptation of the novel described above. Now, your mind does not need to retrieve images and videos from your memory, but instead images and videos are being fed into your brain to become Events and Entities that can be recalled, or referenced, just as previously stored memories are. When reading the book, your mind must retrieve images and videos from your memory, use analogies to make revisions based on the details from the novel, and then create the storyline in your memory as a series of Events. However, with the movie, your mind still needs to retrieve information in order to understand what you are watching, but the videos from the movie can be stored directly as a series of Events. In either case, your mind ends up creating a series of Events based on the story.

You are still probably drinking coffee and eating snacks and putting more wood in the fire, so in both cases you still have a parallel series of Events that are being stored, one fictional and one based on reality. As with the novel, the memories from the movie that become stored need to be tagged as to whether the information is based on fiction or fact. And this point can become very subtle and intricate because for many novels, the background information is true, but the main characters and storyline are fictional. In addition, the story might be biographical, meaning that everything should be based on fact, but the author or director may have taken liberties with the truth in order to make the story appear more interesting. In this case, your mind may have tagged the story as true, but after you find out that some parts are not, you must mentally go into the stored memory and re-tag some of the parts of it.

Creative-Imagination Memory

Two of the main purposes of our imagination are to envision Entities and Events that might be, and to assist with analogies. We use our imagination for the future, past, and present. We use it to predict and make plans for the future, based on our current knowledge and the use of logic. For example, if you are going to a hardware store to buy some lumbar that you will carry home in your pickup truck, you will likely try to visualize if it will fit, and if so, the best way to load it so that it not only fits, but is also secure.

For the past, we will likely think about what occurred during the last day or so, and evaluate whether we responded correctly, or how we might respond differently to a similar situation in the future.

With each of these examples, our minds will store a record of what we thought about, allowing us to later think back on how we resolved each of these concerns, and allow us to make use of our learning either when we get to the hardware store, or when the recent experience is repeated in some fashion.

And we also use our imagination in the present to help us navigate our surroundings. Let's consider an example in which you go into your garage to assemble a new piece of equipment. As you enter, you notice that there is a power tool lying precariously on the shelf, and your mind imagines it falling onto the floor and breaking, or falling onto something else and breaking it, or falling on your head. Your mind will likely be retrieving one or more memories of an object in a similar setting that was falling, and, using analogy, your mind replaces the remembered Entity with the current power tool. And, since the surroundings were likely different, this too is being replaced with the current surroundings. So, you are mentally predicting the power tool falling, and what it would look like, and what would happen if it did.

Your mind then must decide what to do with the power tool. Maybe there is enough room, and you can just slide the tool further onto the shelf. The main storage cabinet for this tool is all of the way on the other side of the garage, but you don't want to put it away yet. There is another shelf closer to where you are going, but it does not have much room.

The point here is that you need to determine your options and then evaluate each option based on the significant criteria, which could be one or two things or many, and may include memories of positive or negative Events related to this tool or one of the locations. You need to give weight to each criterion, which might include distance, available space, effort, safety, and convenience, and you must determine the importance of each. You then decide as to the best option (course of action) based on a quick evaluation of the relative importance of the criteria.

After moving the tool to a better location, you continue over to your workbench, which has the equipment that you recently purchased. You look at the box to see how to open it. There is tape across the two main cardboard flaps and along the side of each flap, and you know that you need something with a sharp blade to cut the tape. Two Entities pop into your mind: your pocketknife in your pants pocket and the box cutter in your toolbox, several feet away. As you instinctively reach for your pocketknife, since it is closer, your imagination visualizes the knife in your hand and the blade slicing through the tape and running the full length of the box. But you are not sure how deep that you can cut without scratching the device in the box, and your mind imagines doing just that. Then you recall the box cutter, and you imagine sliding the blade in and out and that there are settings that allow you to adjust how far the blade sticks out. Then you visualize cutting the tape with the blade at a shallow setting and not scratching anything inside the box. You then walk to your toolbox, open the drawer, grab the cutter, walk back to the box, adjust the blade to a shallow setting just as you had imagined, and proceed to cut the tape, exactly as you had visualized.

The flaps of the box are folded out of the way, and the various parts are carefully removed from the box, and each inspected, as it is lifted out. You locate the instruction sheet and study the assembly procedure. Your mind is likely now using your imagination to mentally move the pieces together, rotating them as necessary as they approach each other.

You are now ready to assemble the first two parts, which require the pieces to be aligned correctly and the designated bolt to be slid through the adjacent holes from a specific direction, after which a bolt is screwed on with a clockwise rotational motion. After mentally rehearsing the task, you pick up the two parts and move them toward each other and line them up, as you have just imagined. You locate the proper bolt, nut, and washer. You mentally move the bolt to the correct location, simultaneously rotating it so that it aligns properly with the hole, after which you actually pick up the bolt and implement what you just imagined. With some gentle wiggling, the bolt slides through the hole until the bolt head touches the part and the threads are showing on the other side. Next, a similar activity is carried out with the nut to pick it up, move and rotate it into position, then rotate it clockwise to screw it onto the bolt. Finally, you will use two wrenches to tighten the bolt, which required its own version of imagination and implementation.

As you carried out each step of this example, you were also creating a series of Events for these activities that become stored in your memory. So, in addition to the Personal Environs memory of the garage and its contained Entities, and your Self-Based memory of your body's sensations and interaction with the garage Entities, you also have stored in your Creative-Imagination memory, in parallel, the series of Events in which you imagined and implemented the assembly of the equipment. Tomorrow, or at a later time, when you remember thinking through how to assemble this piece of equipment and then doing it, you will also recall how you felt, where in the garage you assembled it, that you had to first move the power tool on the shelf, and about what time it was.

I don't know if your mind creates parallel tracks of different types of memory exactly as I just described. But I do know that your mind remembers the type of information discussed in the example. Whether or not the human brain organizes memory in this way, it seems to be a logical way to do it, and I believe that it would make sense for this concept to be implemented in an AI with AGI. This would allow the AI to organize the complexities of its surroundings and its activities within these surroundings in a way that simplifies the memorization process while incorporating all of the important details.

However, there are some differences that could be implemented in an AI system that take advantage of its much-advanced ability to store information, as compared with a human brain. The human brain can store a limited amount of information in its iconic, or short-term, memory, and a small part of this becomes stored as a long-term memory. So, for a complex scene, you may need to see it several times before you can remember all, or at least most, of it. After your brain has converted what it can to long-term memory, the information is wiped from the short-term memory. And the associated location and time information that is associated with the scene may not be stored at the same level of detail in which it was available.

An AI could permanently store all of the visual, and all other sensory information, along with the exact time, and location data in the form of both GPS and other types of location information, such as the name of the room, building, city, state, etc. All of this data could be recorded with a much higher resolution that can be done by the human brain. And all of the sensory information would exist permanently to allow nearly unlimited associations to be made now and at times in the future.

Rules

As discussed earlier, the human brain uses Rules in several ways, some of which were discussed in the chapter on Logic, Reasoning, and Imagination, that represent relationships between Entities and Events. For example, when a pattern is detected, Rules can be created that represent this pattern. If there is a series of numbers, and each consecutive number increases by 2, or is double the previous number, then a Rule can be created representing this relationship. Notice that this Rule can then be used to predict the next number in this series.

For causation, if every time a certain type of Event occurs, another type of Event follows, a Rule can be created for this relationship. The next time that the first Event occurs, this Rule can then be used to predict that the second Event will follow.

These Rules must be stored in memory in a way that they can be located and retrieved. Links evidently are made to the Entities and Events that were used to create the Rule, and each time the Rule is applied to a new situation, new links would be formed. But analogies could be made that might allow this Rule to be applied to a somewhat different set of Events. When these links are created, then each time that a given Event or Entity is thought of, the link would make reference to the Rule and the possibility of the Rule being applied.

Memory and Rule Agents

So how does your mind accomplish the task of organizing all of this information as it arrives in your brain? This is only partially known. But let's remember that our main goal is not to fully understand how a human brain operates, but rather to create an AI with AGI. Since we are trying to develop something (AI with AGI) that mimics what the human brain can do, it makes sense to understand the brain as well as possible. But we want the AI to do essentially what the brain does, even if not in the same way.

I am proposing that for an AI, there be a system of Agents that accomplish the tasks of storing in memory the new information and creating the links between the new and existing information. Think of these Memory Agents as software subroutines that each have specific tasks. They add new information and create links, and by doing so, continually change the mind of the AI. But the Memory Agents themselves do not change. They each have a job to do that always remains the same.

Let's review the bedroom example from the section on Real-Time Personal Environs Memory to understand how this might work. As the AI enters the bedroom, let's call it Robbie, the first thing that its mind must do is to determine whether it has been there before, or if this is a new location. An Agent would compare the view of the bedroom with the images and videos of locations in its memory. Note that analogy must be used to make this comparison because the current view probably would not exactly match any of the stored images and videos of the bedroom, even if this room is in its memory, due to the exact viewing angle. Various levels of analogy could be applied, such as determining if the current view is similar to unrelated bedrooms. In this case, the main Entities, their locations relative to each other, and the size of the room should agree. If a match cannot be found, indicating that this is the first time in this room, another Agent much create a new mental construct for this new room. Otherwise, the current images and videos will be added as links to the existing mental construct of the room.

Another Memory Agent would be storing the images as they enter Robbie's mind, and yet another Agent would be creating new mental constructs for the individual Entities or creating a link from the current images and video to the mental construct of the bedroom. Other Memory Agents are making links between the individual Entities (of the bedroom) for the current and existing images and videos. These links allow other Agents to use the mental constructs of the Entities to identify what is currently being observed by Robbie.

Other Memory Agents would be used to retrieve memories, as needed. And still other Agents would search for associations within this and previous memories to form Groups from individual Entities.

Rule Agents would search for relationships and create logic Rules, while other Rule Agents would search for opportunities to apply Rules, and then actually apply them. Using Memory and Rule Agents, which do not change, in themselves, would enable the brain of the AI to change by adding incoming information in an organized fashion, and making use of that information in an intelligent way.

11

Choices

"A customer can have a car painted any color he wants as long as it's black"

- Henry Ford

Whenever we buy something, or decide to do some activity, we usually have a set of criteria to help us make the correct selection. The list of must-haves and nice-to-haves could be very detailed or maybe it just covers the main points. And when we start searching, we always hope that we can find a product or activity that perfectly matches all of our requirements and desires. But, as we all know, this almost never happens.

There are only so many different cars, houses, shoes, tools, cruise ships, movies, restaurants, etc. So, the task becomes one of determining what our options are, or choices, and determining which selection best suits our criteria. But, of course, it is often difficult to decide how to prioritize our criteria list. Even after we have divided our list into needs and wants, it is difficult to prioritize the points on each of these lists.

Let's consider the purchase of a new car. Here is a typical list of important considerations:

- Timing
- Cost
- Style
- Color
- Brand
- Model
- Size
- Range
- Horsepower
- Electronic Accessories

- Physical Accessories (nice wheels or leather seats)

From the overall list, we can select the "needs," which might include:

- Timing
- Cost
- Style
- Size
- Range
- Specific Electronics

The remainder will be on the "Wants" list. We then need to prioritize the items on each list. For each person, each of these lists and their priorities will be different, and depend on their life situation, finances, and personal preferences.

We would have probably already done some initial research either formally or through casual observations, but now, additional exploration is needed. This could be internet searches, inquiries to friends, and visits to auto dealerships for test drives. New information gathered will likely be used to refine the lists.

At some point, we have to get serious about making the decision. We really like Model D, which fits most of our criteria best, but the cost is higher than we really want to pay. Could we live with Model E or F, which are within our budget, but the range is less than desired on the one, and the other does not have quite the right electronics. We probably need to re-evaluate the importance of these criteria relative to their additional cost.

And the choice might not even involve the cost. Our two favorite cars could be two that cost about the same, but each have a different set of desirable features.

Every time we choose to purchase something or engage in some activity, we need to go through a similar process. However, the importance of the decision depends on the cost, lifecycle, and other factors. Our choice will be less important for which meal to select at a restaurant compared with which house to purchase.

For most things, our only option is to make a selection between existing choices. We are not going to create our own car, make our own cell phone, or build our own cruise ship. And we probably will not be able to, or want to, make our own movie or form our own orchestra. And island vacations are pretty much limited to the islands already in existence.

However, there are a few cases in which we can get much closer to the exact object that we want. Imagine that you have searched for a specific bookshelf to fit in a difficult location in your house, but none of the ones that you found were quite right. However, you are an excellent woodworker, with the skills to build almost anything from wood.

So, you sit down for a few hours and draw your bookshelf in exacting detail that matches every one of your preferences with regard to size, shape, shelves, and decorative accents. At this point, you need to think about exactly how to build the bookshelf, and you need to purchase the wood and hardware required.

But now you need to choose which type of wood from the limited species of wood available. You need to select the brand and specific color of the stain, as well as the polyurethane. You must decide whether to use nails or screws, and what size. And when you are ready to start building, you need to choose how to measure each of the pieces of wood, which saw to use to make the cuts, and which paint brushes to use to apply the stain and polyurethane.

So, now we are able to get exactly what we want, without having to determine which selection criteria are more important than others. But we have just moved the "choices" requirement to a different level. At the construction level of the bookshelf, we have to make numerous selections from limited choices, with the likely dilemma that most of these choices are not perfect, and for each choice, we need to make the one that best suits the situation.

For almost every decision that we make every day, we need to have formulated what our ideal choice would be, then select the best option from the available possibilities. And the effort involved should be commensurate with the importance of the decision.

An AI with AGI capabilities should understand these concepts and make decisions on a similar basis as a human would. However, with the correct decision-making process, an AI could make better, and often much better, decisions than a human because the AI could acquire, and make use of, many more details. And it should often be able to evaluate all of the criteria in a more logical fashion than a human because it could weigh each point more precisely.

12

Predictions, Planning, Execution, and Evaluation

"It is difficult to make predictions...particularly about the future."

- Mark Twain

As we go through each day, much of what we do mentally is to try to predict things. What will the weather be like? How heavy will the traffic be on the way to work? What will the results be if I run this experiment? What will my colleagues think about my presentation next Tuesday?

But predictions also involve the past. When something adverse occurs, we want to understand what happened to cause this Event to occur. This is also a form of prediction, since we are trying to understand what Event could have occurred first, that if we knew that first Event did occur, we might have predicted that it would have caused the negative second Event (that did occur) to have occurred.

The following four concepts tend to summarize how we use prediction with cause-and-effect:

1. Understanding what happened in the past
    - (the Effect is known, and we want to determine the Cause)
2. Speculating about what could have happened in the past
    - (the Cause is known, and we want to determine the Effect)
3. Determining how to make something happen in the future
    - (the Effect is known, and we want to determine the Cause)
4. Predicting what will happen in the future
    - (the Cause is known, and we want to determine the Effect)

The first two are about the past, whereas the second two refer to the future. However, there are two different starting points involved here. The first concept, understanding what happened (in the past), really means knowing about an Event (Event 2 - Effect) that occurred, and trying to determine what Event (Event 1 - Cause) must have occurred first that caused Event 2 to occur. However, for Concept 2, we are speculating about a possible different first Event (Event 1 – Cause) that could have occurred (although it did not), and what second Event (Event 2 – Effect) might have been caused to happen by this possible first Event.

So, there are two different starting points because for the first concept we know the effect but are trying to determine the cause. But for the second concept, we know the cause (or what could have been a cause) and want to predict the effect (that would have occurred if the cause had actually happened).

The third and fourth concepts are analogous but refer to the future instead of the past. When we are trying to determine how to make something happen (third concept), it means that we know the result we desire (Effect), and we want to determine what activity, or series of Events (Cause), would result in the desired outcome (Effect). On the other hand, for the fourth concept, we are aware of something that is happening, or will happen (Cause), and we are trying to predict what Event (Effect) will likely occur in response.

So, for all four concepts, which involve cause and effect, it's a matter of whether the unknown is the input or the output, and whether the Events occurred in the past or will occur in the future. I need to point out that I have oversimplified these concepts for the sake of understanding. Many cause and effect relationships involve more than one cause and/or more than one effect and there may be a series of Events that represent the cause. But the main idea of the four concepts still applies.

To make predictions for each of the four concepts, we must rely on cause-and-effect relationships that we are already aware of and make analogies to the new situation. For example, if we have dropped one or more water glasses on a ceramic floor, we would be well-qualified to predict what would happen to another water glass, even if its color, shape, and the decorative pattern on the side were all different than the ones that had been dropped.

The description above, and predictions in general, tend to refer to causation related to Events. But predictions can also pertain to patterns and to Entities or Groups. For example, if we have a series of six numbers, in which each consecutive number is 2 times the previous number, we could use this pattern to predict the seventh number. And if we have a 1000-piece jigsaw puzzle with a gap in the middle, we could make use of the surrounding images to predict what the pieces must look like in the missing section. Or, if we determine, after several weeks or months, that the traffic to work is the busiest every day at around 7:00 AM, we can likely predict that tomorrow and the next day will be the same.

So, we can extend our summary list to include concepts related to patterns:

- Determine missing Entities in a pattern (interpolation)
- Predict the extension of a pattern (extrapolation)

A large fraction of our mental activity is focused toward understanding what happened in the past, or what could have happened under different circumstances. This helps us learn from the Events we are involved in, Events involving others (including other living things), and Events that we are aware of, so that we can create cause-and-effect relationships, and we will know how to repeat desirable experiences and avoid those that are unpleasant. And we can use analogy to extend these learnings to other situations.

It is important to point out how most of what we do is based on something that we learned previously and the use of analogy. We are usually either repeating a previous experience, following directions or a procedure (recipe), or applying something that we learned from a class, a book or journal, the internet. And since things are never exactly the same, and sometimes very different, we need to use analogy to adapt what we know to the new situation.

Predicting What Will Happen in the Future

With regard to the future, if you observe an Event that just happened or is happening, it is likely you will try to predict what else is liable to occur as a result. If you're a skier that lives by a ski resort, and you see that it is snowing when you get up in the morning, you'll probably have the urge to head up to the mountain because you would naturally predict that there will be some nice powder on your favorite run. If, instead, you're on a cruise ship, and one evening all the normally empty slots around the ship are filled with sick sacks (barf bags), even though the sea is calm, you're likely to predict that the ship is headed into a storm and that you will need some seasick medication.

Notice that for these situations, we have no control over the outcome. We are simply observing a cause and making use of all of the knowledge we have in the form of cause-and-effect relationships to predict the effect. It could be good, bad, or indifferent for us, but we know that the result will not depend on what we want. We may be able to act on our prediction to take advantage of, or avoid, the expected outcome. But we are not going to be able to change the weather forecast or height of the waves.

We predict the future hundreds of times per day, based on our years of informal experience and formal education. These predictions are based on memory/experience, analogy, causation, patterns, math, and imagination. This topic has been discussed earlier in the book, but I would like to highlight some of the examples that we experience every day, either as a layman or a professional in some field. Here are some examples of actions that we take (cause) and the result that we predict will happen (effect):

- Flip a light switch ➔ the light comes on
- Increase the thermostat setting ➔ the room gets warmer
- Turn on the burner for the stove ➔ the burner gets hot
- Heat water in a pan on the burner ➔ the water boils
- Put an ice cube tray with water in the freezer ➔ the water turns to ice
- Turn the ignition key on in the car ➔ the engine starts
- Put the transmission in drive and press down on the accelerator ➔ the car goes forward
- Turn the steering wheel (while the car is moving) ➔ the car turns in that direction
- A construction worker mixes cement with the right amount of water ➔ the solution slowly turns to solid concrete

- The pilot of a 747 pulls back on the yoke at 190 MPH ➔ the plane lifts off the runway
- An anesthesiologist administers the right amount of a drug to patient ➔ the patient goes to sleep
- A chemist mixes acetic acid, ethanol, and catalyst at a specific temperature ➔ ethyl acetate is formed

We, including the professionals, know about these relationships from personal experience and training, and can now make predictions based on them. You know that when you flip the light switch on that the light WILL turn on and turning up the thermostat WILL make the room warmer. And this list barely brushes the surface. We each are aware of many thousands of these relationships from the minutia of a light switch to the life-or-death skills of piloting a 747.

Analogy plays an important role in each of these cause-and-effect relationships. There are many types of light switches, with several styles for wall switches, but some of the switches are built into lamps, or in cords, and they vary in color. If you enter a house that you have never been inside of, and it has a different type and color of light switch on its walls, you will probably figure out immediately how to manipulate the switch to turn on the corresponding light. You are able to do this by using analogy. You know that the color is not important, and that wall switches tend to be about a certain height from the floor, and that they usually toggle back and forth in some way. If the switch turns out to be a slider, and you have never seen one of these, it may take another second or two for you to determine how to move it to get it to work.

And this is the same for almost everything. Thermostats come in many types and styles, there are many different types of knobs for stoves, and car accelerators require different amounts of pressure. But for each of these devices, they have certain important Attributes in common, and we can adapt, within a certain range, to the differences.

But sometimes, the difference between the old and the new is too much, and you need to be shown or relearn the cause-and-effect relationship for the new device. An example of this is the push-button starter, or keyless ignition, on cars that started to appear in the later 1990's. Before that time all cars had a key that was placed into a slot and had to be turned clockwise to make the engine start, whereas you press a button with the keyless ignition. If your first exposure to this was with a rental car with no pre-warning, you probably tried a few times to get the key to flip out of the key fob, and maybe even searched in vain on the dashboard or steering column for the key slot, before going to ask for help.

Things are constantly changing beyond what analogy can accommodate, which requires additional training, research, or asking someone for help. Examples are software, tools, technical instruments, new drugs for medical professionals, and instrument and control changes on aircraft, just to name a few. But even after learning about the new software, equipment, drugs, etc., we still will need to rely on analogy to cover for the inevitable differences between exactly what we learned and the real-life experience of making use of this new knowledge.

The value of patterns for making predictions was discussed at the beginning of the chapter. The application of patterns also requires the use of analogy because there will always be some differences between the pattern that you experienced and the current situation.

Imagination also gets involved because you are usually envisioning everything before you interact with it. You have a mental image of where to look for a light switch, you envision inserting the key into the ignition and turning the key before you actually do it, and the pilot anticipates a certain indication on the airspeed indicator and other instruments, and himself pulling back on the yoke at just the right time. And before using a power tool, you envision what the tool will do before turning it on. If you are trying to assemble a piece of equipment, you imagine moving the parts together, maybe rotating them or adjusting the angles, and how the bolts or other fasteners will be moved into position and screwed together. And what type of tool will be needed to tighten the fasteners.

Determining How to Make Something Happen in the Future

One of the most important uses of prediction abilities is to help create or do something. For example, if you want to bake a cake, create a recipe or procedure that can be used to bake a cake, get groceries, build a cabinet, run an experiment in the lab, or go on a vacation, your mind has envisioned some end result, which may or may not be tangible, and you need to determine the steps required to achieve your goal.

The following discussion is based on how a human would make something happen, but the overall goal is to understand the thought process that would be required for an AI to do this. Since an AGI should be able to perform at least as well as a human at this task (and on many other tasks), I am using the thought process of a human as a model for an AGI.

The goal might be to create a recipe or procedure, rather than an actual product, which could be used later by you, or by others, to make a product. When initially creating the recipe, you need to predict what will happen with each step (Event) that is carried out, and there will probably be a lot of trial and error involved. The value of a very detailed recipe is that all of the trial and error has been done before, and you do not need to worry about repeating these experiments. However, even with a very detailed recipe, there will still be some predictions, and possibly some trial and error, involved, because you will not be able to exactly replicate the conditions used to make the original recipe, and the use of analogy will be required. The equipment may not be the same or the raw material might be slightly different.

There are probably at least five levels of complexity, regarding predictions, for making or creating something:

- Follow a recipe or procedure to create something as close as possible to the original
- Basically, follow the recipe or procedure, but make some modifications. For making bread, change the type of flour, the size of the loaf, add nuts, onions, peppers, or fruit, or change the cooking time.
- Adapt your skills to a different kind product, but still within same general area. For example, use your cooking skills making bread to now make cookies, or make a wooden table when your experience is with making bedroom dressers, or as a chemist make ethyl acetate based on your knowledge of how to make methyl acetate.
- The main variables are known that could affect the creation of a product, but it is not known which of these variables are important. A new chemical reaction could be an example of this type of activity. Temperature, pressure, concentration, and

reaction time could all play a role, and usually a statistically determined designed set of experiments is carried out to determine the cause-and-effect relationships, which can then be used to predict the results of future experiments.

- Adapt knowledge from one field to a completely different field. Some examples:
    - In 1945, Percy Spencer was working with a radar set, which used a magnetron tube, and found that a candy bar in his pocket had melted. This led to the development of microwave ovens. [43]
    - The Wright brothers studied bird wings, and their flight patterns, as the basis for designing the wing for their first airplane. They also adopted the concept of the boat propellor to design the propellor for their airplane. [44]
    - The development of the PID (proportional-integral-derivative) controller, used extensively in many applications, was directly inspired by the observations of Nicolas Minorsky watching ship captains maneuvering large vessels in the 1920's. [45]

Needs are the Incentives for Goals

But what is the driver behind the creation of goals, whether they are used to create a product or to accomplish something? All goals are the result of having a need or desire that we want to satisfy, as was discussed in the chapter on Needs and Emotions. For a human, the type of need might be one of the following (although there could be others):

- Respond to a request from someone, such as to provide information to, or do something for, someone
- Add to current knowledge
- Satisfy a biological need
- Do or make something that is pleasing to your senses (looks, sounds, feels, smells, tastes good)
- Preventative maintenance (to avoid future problems)

For an AI, most of these needs could still apply, except maybe for the third and fourth ones. But an AI could have needs that are analogous to biological needs. For example, a mobile robot (maybe humanoid) would need to charge itself periodically, so it would likely have a desire to get to a charging station when its battery is low. It should also have a desire not to damage itself, and if it does become damaged, to take action to have it repaired. For the fourth type of need, an AI with AGI could possibly be programmed so that it experiences a form of pleasure in seeing beautiful objects or hearing good music. I could envision an AI employed as an art or music critic, or even as an artist or composer (although I hope that we leave these occupations to humans, but an AI might act as an assistant to identify performers or works of art that a human could then evaluate). Experiencing one of the needs listed above is the incentive for creating a goal.

Goals

A goal is essentially a desired state of existence that we would like to achieve for ourselves, or an Entity (maybe a product) or Group which may or may not already exist. We need to use our imagination to envision ourselves, or this Entity or Group, in this desired state with a set of Attributes that represent this state. Whereas when we normally discuss Entities, we are referring to something that exists, or did exist, with a specific set of Attributes. But the Attributes for an Entity that represents the Goal are essentially metrics that we would like to achieve. For most Goals, we will not reach the exact Attributes that we desire, but we want to get as close as possible within certain acceptable tolerances.

SubGoals and SubEvents

To achieve the Goal, we will need to start from some Initial State and implement changes through what we will call the Goal MultEvent, which is a series and/or parallel set of Goal SubEvents, as shown in the next figure. However, we may not be able to define the Initial State until we determine the details of the MultEvent. For example, if we are building a piece of furniture, the exact boards and fasteners that we use may depend on how we plan to build it. Figure 20 indicates that there may be multiple Initial States.

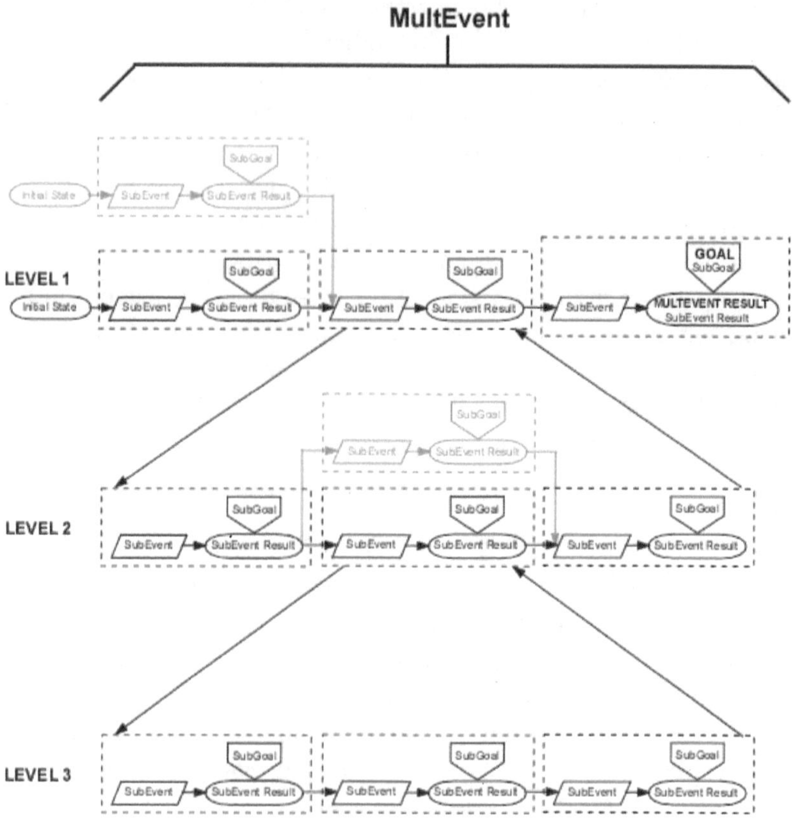

Figure 20. MultEvent with hierarchical levels.

The Goal SubEvents are slightly different than we normally think of Events because each one has to achieve a specific SubEvent Result, based on a SubGoal, with their sets of Attributes. Imagine that we are building a bedroom dresser cabinet with drawers. Each drawer would need to be made to near-exact specifications, which are based on the Attributes of the associated SubGoal, in order that the drawer fits perfectly into the dresser cabinet.

In addition, the construction of each drawer requires its own set of SubEvents and SubGoals, which would correspond to Level 2 on the figure (Level 1 is the overall project). For example, each of the sides, the back, front, and bottom would need to be built to exact specifications so that they will all fit together to make a drawer that will fit into the dresser. And each of these pieces might be made from smaller pieces (Level 3 with a deeper layer of SubEvents and SubGoals). So, the SubEvents and associated SubGoals can be quite hierarchical. In addition, since each drawer can be built independently of the other drawers, as well as the dresser cabinet, these SubEvents can be carried out in parallel or series, as indicated on the figure. This same point also applies to the parts for each drawer.

Each of the Goal SubEvents will have a set of Attributes, which would include at least the following categories:

- Entity
- Action
- Time
  - Start Time
  - Elapsed Time
- Cost
- Probability of the SubEvent succeeding
- Safety

These Attributes will be used to compare one possibility or option for the SubEvent with another SubEvent possibility or option to accomplish a given SubGoal. This point will be discussed in more detail below under the section on MultEvent and SubEvent options.

Goal and SubGoal Attributes

The Attributes for the Goal are very similar to the Attributes for Entities discussed in the chapter on Entities. However, some Attributes will be more important than others, depending on the type of Goal. These Attributes might include:

- Size
- Mass
- Materials of Construction
- Color
- Cost
- Time Required
- Effort

- Importance
- Availability of resources
- Quality-Adequacy
- Probability of Success
    - Whether or not this Goal can be accomplished
- Safety
- Warning/Encouragement from someone
    - Technical reasons why this Goal is good or bad
        - "John made a cake using the same cake mix and he said it was delicious!"

The SubGoal Attributes would be similar, but there would be differences. For example, for making a bedroom dresser, the size, mass, time required, and effort for any of the drawers would be lower values than for those of the entire dresser.

MultiEvent and SubEvent Results

A MultEvent Result is the actual state that is created due to the MultEvent, and its Attributes must be compared with those of the Goal. This also applies with the Attributes of SubEvent Results and SubGoals. This point will be discussed in more detail below.

Goal Priority

The Goal Priority represents the importance of this Goal relative to other Goals that we have. If we are baking a cake and one of our other Goals is to buy more cake mix next time we go into town, making the cake now would likely have a higher priority. However, if the other Goal is, instead, to get out of the house because the kitchen is on fire, the Goal to make the cake would hopefully have a much lower priority.

The Goal priority is based on the consideration of what will happen if this Goal is not carried out and what will happen if it is. But there is also a time element, which could be divided as follows:

- Now
- Short-term (30 minutes to a few hours)
- Mid-term (a day to several weeks)
- Long-term (a month to several years)

So, there would be a multi-faceted importance factor assigned to the Goal that is time-based, and the outcome of accomplishing, or not, the Goal can also be categorized as a type of need based on a scale similar to that presented graphically as a pyramid by Maslow in 1970 (see the figure in the chapter on Needs and Emotions).

Here are some examples of how these concepts would translate into priorities. You might be kind of hungry, and your Goal is to eat lunch. The consequence of not eating Now would be a slight amount of physical discomfort (hunger). The consequence of not eating in the Short-term would be extreme discomfort. Not eating in the Mid-term means that you experience extreme hunger, pain, and then you die. Hunger is a physiological need, which is on the bottom level of Maslow's pyramid, indicating it as a primary need. So, the Goal of eating would, in general, have a relatively high priority, but it has a time element. The need and priority might be only moderate now, but would increase as time progresses.

As indicated, Goals and their priorities are dynamic, as was emphasized by Marcus and Davis [46]

Let's compare with some other Goals. Maybe you are working on a project, it is nearly lunchtime, and you are getting hungry, as just described. For the project, you may need to glue two pieces of wood together, which will only require a few minutes, the glue will take about an hour to dry, and you cannot proceed with the project until the glue is dry. If you wait until after lunch to glue it, the project will be delayed by an hour, which could be translated into the loss of income, if this is for work. A comparison could be made as to how much priority of eating will increase in the few minutes that is required to glue the pieces, compared with the increase of priority for gluing if lunch is eaten first. This situation would likely raise the priority of gluing the boards above that of eating immediately.

Goals with a low priority should be delayed relative to higher priority Goals. However, the priorities of SubGoals would also be compared with the SubGoals within the same Goal if the SubGoals are parallel, or with the SubGoals of a different Goal.

An example of parallel SubGoals would be the construction of the drawers discussed previously. The assembly of the drawers should be parallel tasks, but there might be a reason to construct them in a certain order, or at least one of them before the others. For the SubGoals of different Goals, an example might be that you are constructing one piece of furniture, but before beginning work toward one of its Sub-Goals, you stop to glue two boards together for a different piece of furniture for a different Goal.

Research

Prior to determining the steps involved to accomplish the Goal, it is important to conduct at least some research. Let's return to the dresser example. What tools and work area are available? If you need something, what would you get and where would you buy or borrow it? What boards and fasteners do you have? Do you have the expertise for each part of the project? If you are lacking in a skill, could you learn about it from books, trade journals, internet, on-line videos, or talking to a friend?

Knowing this information will help to decide which options are available for construction, in other words, what types of SubEvents does it make sense to consider. This aspect of the project will likely be somewhat iterative in that after the initial research is conducted, additional research may be required as the construction options are identified.

MultEvent and SubEvent Options

Once we know what the Goal is and we have researched our resources and enhanced our knowledge about the project, we need to determine how to accomplish the Goal. What options for steps, or SubEvents, can we conceive that will allow us to go from an Initial State (or multiple Initial States if several Entities converge) and reach the Goal? This involves three main activities:

- Create Options
- Evaluate Options
- Select Option

These activities apply not only to the overall MultEvent of the Goal, but also to the SubEvents for the SubGoals, as will be discussed next.

Create Options

At least one set of SubEvents will need to be created, which will comprise a MultEvent, that accomplishes the Goal. However, normally, there is more than one way to do something, so other MultEvents will likely be created and considered.

For many of the SubEvents of each MultEvent, there will be a hierarchically lower level and set of SubEvents for the associated lower level SubGoals. Consider the bedroom dresser example from the beginning of this chapter. There are the SubEvents to make the dresser, the lower-level SubEvents to make each of the drawers, and an even lower level to make the each of the sides and the bottom of each of the drawers.

There are several starting points for creating SubEvents that will lead toward the Goal. However, it is important that a person, or an AI, has significant experience with how Entities act and interact with other Entities (as most people do) in order to have a significant mental storehouse of cause-and-effect relationships, mental videos and images, and experience with assembling and disassembling Entities and Groups, and moving, lifting, throwing, stacking, dropping, breaking, repairing, washing, painting, and observing them. And experience is needed with following directions, and relating maps, models, and drawings to real locations and Entities. A person gains this knowledge as they interact with the world beginning with childhood, and continues with formalized training in school, college, or trade school. An AI would either need to have similar experiences as an embodied Entity, or this information would need to be added to its memory database so that it understands these concepts without having personally experienced them.

This knowledge forms the basis for creating SubEvents to accomplish something. In almost all cases, the knowledge from previous experiences will not work directly as one of the SubEvents for a new Goal because it would not exactly match the current situation. But usually, analogy and imagination can be used to adapt one or a few of these memories to serve as an adequate SubEvent.

Let's consider the four most likely starting points for creating SubEvents:

- Procedure, recipe, map, model, or drawing,

- Previous experience with a similar project or Goal,
- Reverse engineering an existing Entity that is similar to the one represented by the Goal,
- Envisioning the final product (currently non-existent) and mentally working backward to determine the required steps or SubEvents (in a way, this is an imagination-based version of reverse engineering).

I will point out that many of the SubEvents for these starting points represent cause-and-effect relationships or patterns.

Continuing with the bedroom dresser analogy, if we have a drawing of how to build the dresser, and/or step-by-step instructions for its construction, as indicated by the first starting point above, then following either of these types of directions simplifies the project somewhat. However, as mentioned above, we need to have experience with relating drawings/instructions to real objects, working with power tools, and working with wood. And it would be helpful if we had previously built a piece of furniture that was somewhat similar, such as a different type of cabinet, in reference to this last point.

But we still need to select the lumber, tools, and work area that we will use. We need to determine the type of wood, where to get the lumber, the exact size boards, and the type of fasteners. We also need to select the tools that we will use, such as saws, screwdrivers, nails, and a hammer. And, of course, we must determine whether to do the construction in the garage or outside. Throughout every aspect just described, analogy and imagination are needed to adapt available knowledge to the current project because history, or previous experiences, never exactly repeat themselves.

If we have access to a bedroom dresser that is very similar to the one we would like to build, we could reverse-engineer it (third point above). We would hopefully be able to pull the drawers out to study the construction techniques and slide mechanism. That would also allow us to look inside to observe the internal construction details. Even if the dresser that we want to build is a different size, uses different wood, has a different number and size of drawers, and different mechanical slides, we can use analogy and imagination to adapt the construction details from the existing dresser to the one that we would like to make.

And even if we are not basing the design of our dresser on an existing one, we can use our imagination to conceptualize the dresser that we would like to build (fourth point above). Then work backward to determine how to build it. Of course, as we imagine it, we should make one or more drawings, with dimensions and any pertinent details, to make sure that our vision is actually what we want and that we are able to build it. We can then base the construction on the drawing, as indicated by Point 1.

If we do not have a procedure, we would need to determine how to divide the project up into logical steps or SubEvents. For one of the drawers, we should be able to make use of our general experience assembling and building other projects to decide that a set of SubEvents would be required for each side of the drawer, the front, and the bottom. Each of those SubEvents would probably have several lower-level SubEvents, such as cutting each side of the board, and possibly gluing two (or more) boards together to make a side, the front, or the bottom. There might be the desire to router the front board for decorative purposes. Then there would be several SubEvents to attach all of the boards together to create the drawer. And there would be final SubEvents to stain or paint the front. Similar logic could be used to divide the construction of the dresser frame into SubEvents.

After a first option is created for the drawer construction, we might want to consider other options, such as for different types of saws that could be used to make the cuts based on how precise the cuts need to be, how the sawdust is ejected, what type of workspace is required, alternate ways to support the boards, different fasteners, etc. And the order of assembly for the boards could be open to debate. At a hierarchical level down, there may be more than one way to create each of the boards.

There would be a SubGoal for each of the drawers, and a lower-level SubGoal for the creation of each of the sides, the bottom, and the front of the drawer. Each of the SubGoals would have a set of Attributes, as discussed above in the section on Goal and SubGoal Attributes.

Several SubEvents would likely be required for each SubGoal, and there would be a corresponding SubEvent Result associated with each SubGoal, which has Attributes that represent the actual product, that are compared with the SubGoal Attributes. In other words, the

SubGoal is a desired set of Attributes to strive for, but the SubEvent Result represents the actual values of the Attributes. Keep in mind, though, that as the MultEvent and its SubEvents are being determined, we will not know with certainty how closely the SubEvent Results will agree with the Goal and SubGoals. We need to use our imaginations, software, or calculations to estimate, as well as possible, the SubEvent Result. It may be possible to avoid actual testing until the entire MultEvent has been envisioned. However, often it is desirable or necessary to actually test one or more SubEvent Results to ensure that it (they) will be adequate before continuing with the creation of the other SubEvents.

Each of the Attributes must be weighted, depending on its importance. The Attributes for the Goal and the SubEvent Results are compared and the differences minimized by adjusting the Attributes of the associated SubEvents. There would need to be tolerances assigned; for example, the tolerance for each dimension might be 1/16".

An important point is that we may have several options for the SubEvents used to achieve each SubGoal, but each SubEvent Result has to agree with the corresponding SubGoal, within acceptable tolerances.

One other point on the creation of SubEvents is that for each one, imagination should be used to mentally implement it to try to make sure that that SubEvent will accomplish what is expected and that there are no fatal flaws. Eventually, the best option for each SubEvent will be tested in reality as the final check, but you, or an AI, want to be careful that you (or an AI) avoid not thoroughly thinking through each option so that you don't reject the actual best option and select an inferior option.

Evaluate Options

Each of the MultEvents options should have resulted in a MultEvent Result that agreed with the Goal within acceptable tolerances. What we are now comparing are the MultEvent options, but we need to look at the Attributes of the SubEvents for each of these options and create a summary of these Attributes for each MultEvent that can be used for comparison.

Shown below are some of the most important SubEvent Attributes, as was previously presented. There would likely be more Attributes, but for simplicity, let's only consider these at this time.

- Entity
- Action
- Time
    - Start Time
    - Elapsed Time
- Effort
- Cost
- Probability of the SubEvent succeeding
- Safety

Time, cost, probability, and safety would usually be the most important Attributes to compare. The Entity and Action cannot be directly compared between SubEvents, but would affect Attributes such as time, cost, probability, and safety, which can be compared.

Each of these Attributes should be weighted and then combined in some fashion for all of the SubEvents of a given MultEvent. For example, the elapsed time for each SubEvent in a MultEvent can be added together to get a total elapsed time, which can be easily compared with the total elapsed times for the other MultEvent options. The probabilities, as fractions between 0 and 1, of all of the SubEvents of a given MultEvent, can be multiplied together to get an overall probability. Safety might also be assigned values between 0 and 1, but more appropriately, there should be sub-levels of safety, such as falling, weight, inhalation, flammability, etc.

Select Option

The best option for each SubEvent for a given MultEvent option would be selected based on the criteria discussed above. Then the best MultEvent option would be selected. However, it is possible that the best one will not be able to be determined before actual testing. So, the Option Evaluation process may only be able to narrow down the number of MultEvents to two or three, rather than just one, prior to MultEvent Testing.

MultEvent Testing

After all of the SubEvents for a given MultEvent option have been determined, a test should be made. For a cake or piece of furniture, this would mean actually making one of these based on the MultEvent procedure. For a car part or refrigerator, it would likely mean making a prototype.

Evaluate Results

The MultEvent Result would be compared with the Goal to determine if there is satisfactory agreement. This will usually mean comparing many, or all, of the SubEvent Results with the corresponding SubGoals.

If, during the testing phase, a problem was found with one of the SubEvents, it may be necessary to revise the SubEvent before continuing with the remainder of the MultEvent. If the entire MultEvent is completed, but there is a problem with the agreement between the MultEvent Result and the Goal, a determination must be made as to which of the SubEvents is the root cause of the problem, and then this must be corrected.

As was discussed above, more than one MultEvent may be implemented. If they are all successful in producing the desired product, then the MultEvent Option Evaluation process can be repeated based on actual, rather than predicted, results.

Example: Making a Cake

Now let's consider the Goal of making a cake. You might have the urge to eat something sweet, after which you need to decide what specific food would best satisfy your desire. Your first thoughts might be a candy bar, cake, ice cream, or berries. Your list gets refined as you realize that you do not have any candy bars, you don't have a cake and you don't feel like baking one at the moment, you have ice cream, but you don't really like the flavor you have, and you like berries, which you have, but not by themselves. So, you decide that you will bake a cake with raspberries. You have a need (or at least a desire) and you have identified a Goal that will hopefully satisfy that need. Now you need to determine the tasks, or SubEvents, that will allow you to achieve that Goal.

However, the Attributes of the Goal need to be defined, since you may not be happy with just any ol' cake with raspberries. These Attributes would likely include the following (and there may be more, but we will stick with these for this example):

- Brand
- Flavor
- Icing or not
- Size
- Shape
- Effort
- Quality
- Cost
- Timing
- Importance
- Probability (of being able to accomplish the Goal)

You may need to adjust the Attributes based available resources or how energetic you are or how much you are willing to spend. You may not have icing available, or you may not have enough cake mix to make the cake as large as you would like. So, you will likely need to prioritize the criteria list, or Attributes.

We will need to define the priority for this Goal. How important is it relative to everything else we have going on. If it's Saturday, you've finished mowing the lawn, and you REALLY want some cake, then it probably has a high priority. But it would need to be compared with the priorities of other Goals.

Now we need to determine the SubEvents for the Goal MultEvent. But before doing so, we need to evaluate where we are starting from and what resources we have, so we need to conduct some research. Do we have a cake mix and other ingredients, which pans and utensils do we have to choose from, which counters in the kitchen are available, and do we have time right now to bake a cake? In this case, it probably just involves searching the kitchen cabinets, our recipe books, and observing the kitchen counters. However, for more complex goals, we may need to search the internet, go to the library, purchase a book, run a simulation, make visual observations, or consult with someone.

Now, we need to determine the SubEvents required to go from our Initial State to the Goal State. For our case of making a cake, we could invent the series of steps, but doing so may involve a lot of trial and error. So, we would prefer to have a recipe, since it would have been tested by someone else and there would be a greater chance of the cake being made correctly (this is the first of four starting points from the section on Creating Options). Most cake mixes have a recipe on the side of the box, which, let's assume, happens to be the case with the mix that we have. However, we may not want to follow the recipe exactly. For example, we are adding raspberries, and we may not have the same size pan as is suggested by the recipe, which could affect the baking temperature or time. But it will likely be much easier to use analogy, cause-and-effect, and imagination to predict what changes are required than to start from scratch with no recipe.

Obviously, each SubEvent needs to move the process in the direction of accomplishing the Goal of making the cake, and so we are making a prediction for each SubEvent, that it will move the process toward the Goal. Each SubEvent is a cause-and-effect relationship that allows you to make a prediction. If you plug in the electric mixer and press the ON button, the blades will rotate. If you turn on the oven and set the temperature, it will become hot. If you put the correct ingredients together, and follow all of the other steps, the cake should turn out as you expect.

The steps are actually very hierarchical because almost every step can be divided into smaller steps, possibly several times. For example, the step of mixing the ingredients can be subdivided into (1) plug in the cord for the mixer, (2) turn on the mixer, and (3) mix the ingredients. Or, with even higher resolution, walk to the mixer, pick up the plug for the cord, move the plug toward an outlet, push the plug into an outlet, pick up the mixer, place its blades in the ingredients, turn on the mixer, mix until the batter is smooth. And for each of these subdivided steps, we are making predictions. If we walk toward the mixer, we will be able to pick up its cord; the cord will be flexible enough to move it toward the outlet; if we press the plug toward the outlet, the plug will slide into the outlet and that it will provide power to the mixer; if we push the ON button for the mixer, the blades will turn; etc. Most of these cause-and-effect relationships are so ingrained in our minds that we don't even think about them as such. We don't have to think to ourselves, "If I press the ON button for the mixer, I wonder what will happen."

Modifications to the Original Recipe

Now, let's look at the modifications to the recipe. The first one is to add raspberries, and we need to decide several points about how to do this (additional research):

- What state should the raspberries be in?
  - Frozen;
  - Thawed;
  - Warmed to room temperature;
  - Heated slightly;
  - With or without the liquid associated with the berries;
  - Drained;
- When to add the raspberries?
  - At the same time as the other ingredients just before mixing;
  - Or after the cake is baked (placing them on top);
  - At a certain clock time, the actual exact time may be based on other activities you are involved in.
- How should the raspberries be added?
  - Pour into the batter directly from bag prior to mixing;
  - Pour into the batter a little at a time while mixing;
  - Pour into a measuring cup first (to get a specific amount) and then into the batter.

Notice that the first question refers to the state of being (an Attribute) of the raspberries. There could be other Attributes that we are concerned about, such as size, level of ripeness, and whether the berries are organic. Each of these Attributes would have their own set of choices that need to be decided.

The second question refers to the time that the SubEvent (adding the raspberries) will occur, relative to other SubEvents or steps.

The third question deals with the Action that occurs with the Event. Notice that this is a SubEvent that is more complex than it first appears, in that it would likely include at least several lower-level SubEvents, such as picking up the bag, picking up scissors, cutting the bag, and pouring the raspberries. There could be other SubEvents, such as pouring the raspberries into a measuring cup, and then pouring the raspberries from the measuring cup into the batter.

This is a very simple example, but shows that even at a basic level, several questions need to be answered, and several options need to be generated and decided upon for each question. To arrive at the best option, the following steps are required for each of the questions above:

- Create Options
- Evaluate Options
- Select Option

To initiate the options, we use knowledge, analogy, cause-and-effect, and imagination to generate a list of possibilities. We may know from previous experience what to do, or we may be able to predict what will happen if we do certain things. And we will almost certainly use our imagination to envision each option and its potential effects. For the first point, your mind would likely retrieve a memory of frozen berries, how cold they will feel, and maybe how some might be stuck together. Or, if you envision thawed berries, you might think about how they will feel soft and there will probably be red water in the bag.

You are actually remembering previous experiences that were similar to what you anticipate will occur, but your mind is adapting it to the current situation. Maybe your most vivid memory was with strawberries, the berry bag might be a different size, or it may have even been in a different kitchen. Notice that analogy is used to adapt, or morph, each memory to the current situation.

After the options have been generated, they must be evaluated against SubEvent Attributes to determine which option is most appropriate, such as with the following Attributes:

- Entity
- Action
- Time

  - Start Time
  - Elapsed Time
- Effort
- Cost
- Probability of the SubEvent succeeding
- Safety

Your imagination would almost certainly be used to envision a given option and the downstream results of this option. Each of the Attributes would need to have weightings associated with it. The best option would then be selected based on which SubEvent for a given SubGoal has the best score.

Finally, the selected option must be integrated into the recipe.

Along with knowledge and logic, imagination would play a key role in initiating, evaluating, and integrating the options. Visualizing each option would help to define it and mentally "see" how it would fit into the existing recipe. For example, for the option of pouring the raspberries first into a cup, then into the batter, we could imagine cutting a corner from the bag to create an opening and visualizing that if the hole is too small, the berries will get blocked, but if it is too big, the berries will flow out of the bag too quickly and some will probably miss the cup and fall onto the kitchen counter. Then, as we visualize picking up the plastic bag of thawed raspberries, we can imagine how to hold the bag. If the hand closest to the opening is in the middle of the bag, we can visualize the end of the bag flopping over and dumping a lot of berries all over the counter, but if our hand is too close to the opening, the bag will sag in the middle and only a few berries will flow out of the hole.

After the option has been selected for the addition of the raspberries to the batter, it is likely that our imagination will insert this SubEvent, or series of SubEvents (raspberry addition) into the recipe so that we visualize it now as one of the steps, located between two of the SubEvents of the original recipe.

Finally, we would follow the revised recipe to make the cake. It would be tested (by eating a piece of it). If it did not turn out as expected, the SubEvents of the recipe would be reviewed to determine which of them required changes, and the recipe would be revised. At some time in the future, the recipe will likely be tested again.

This example shows that a significant amount of thought is required even for apparently simple decisions about tasks. It also shows that the process is hierarchical, and that within each option, several possibilities need to be considered. Depending on the complexity involved, there could be additional levels of decision-making.

Example: Chemical Reaction

Now, let's consider the more complex example of a chemical reaction. For this example, our Goal is to study the reaction of propionic acid with ethanol. This is a well-understood reaction, but let's pretend that we are the first ones to investigate it. Figure 21 shows the two molecules:

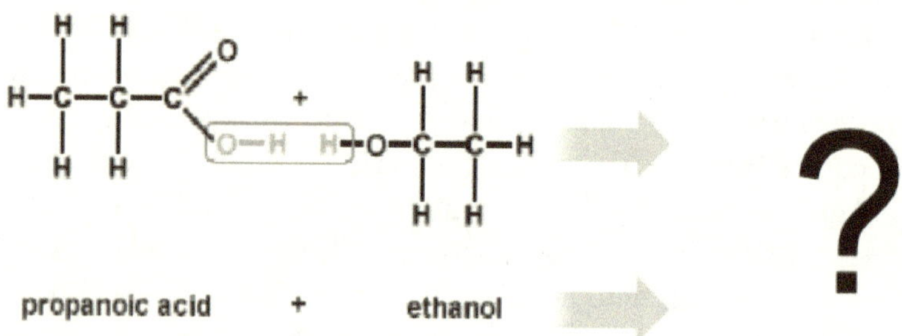

Figure 21. Esterification reaction of propionic acid and ethanol with unknown products.

As with any other activity, in order to accomplish this Goal in any meaningful way, prior knowledge is required, and in this case, in the field of chemistry. Someone with a PhD in chemistry would likely have extensive experience and knowledge of this type of reaction, and almost immediately knows how to proceed. However, a person who has just completed high school chemistry and can at least understand the concepts of molecules and reactions should be able to search for information in books or on the internet to compensate for some of their lack of prior knowledge. In either case, existing knowledge and/or recently acquired knowledge will be used as a basis for this Goal, using analogy to adapt this knowledge. For this example, I will assume that our chemistry knowledge is between that of the high school student and the person with the PhD in chemistry.

Let's begin with the incentive to accomplish this Goal.

Need

The need is probably one of the following:

- Respond to a request from someone, such as to provide information to, or do something for, someone
    - In this case, it might be your boss or a colleague at your company
- Add to current knowledge

Goal

Let's define this Goal with some additional details. Here is the initial question:

- If propionic acid and ethanol are reacted together, what happens?

This question can be changed slightly to the form of a cause-and-effect statement:

- If propionic acid and ethanol are reacted together, ______.

This statement has a cause but is missing the effect. If we can determine the effect, we will have developed a new cause-and-effect statement, which, essentially, is the Goal. This Goal is related to all four concepts about prediction discussed at the beginning of this chapter. We know the cause and we want to determine the effect, as for the second and fourth concepts. However, we do not yet know the cause-and-effect relationship that would allow us to predict the effect. The Goal is actually to determine this relationship so that we can fill in the blank in the statement above. If we can do this, then we could use prediction with any of the four concepts, so we could make predictions about his reaction for the past or future.

For example, if we know the products for this reaction, and therefore the full cause-and-effect relationship, then if we want to make these products, we could use this relationship to predict that we need to react propionic acid and ethanol together (third concept). Or, if we know that propionic acid and ethanol are being combined, then we would be able to predict what products will be made.

However, this cause-and-effect relationship is more complicated than "if I drop this glass on the floor, the glass will shatter." We know from our general knowledge of chemical reactions that the following questions should be asked:

- Does any reaction occur?
- What are the products?
- What temperature is required?
- What are the boiling points of the reactants?
- How fast is the reaction?
- Is a catalyst required?
- How far does the reaction proceed (100% completion?)?

We would have needed fairly extensive training in chemistry in order to know to ask these questions. We could have seen a specific list in a chemistry book or online, or we might have experience with several different types of reactions and have been able to generalize as to what is important about chemical reactions, forming the basis for these questions.

Now let's think about the cause Event and the effect Event. The cause Event involves combining the reactants (propionic acid and ethanol) at some temperature, possibly with a catalyst. For the effect Event, the reaction occurs (hopefully), and the reactants convert to the products (unknown at this point) over some period of time, which could be fast or slow. We will need to determine how to create the conditions for the cause Event, and how to monitor the reaction as it proceeds. However, we need to first define the Goal more precisely.

The Goal could be very elaborate and cover a wide range of conditions, or it could just involve one specific set of conditions. This would depend on the need that was discussed earlier. To keep this example relatively simple, let's assume that the Goal involves the question of what happens in one experiment at one set of conditions. But what conditions should be used? To answer this question, we need to do some research to try to determine the products and to estimate how this reaction might proceed.

Research

We essentially now have another Goal, which is to learn what information already exists about this reaction and the chemical components involved. To do this, we first need to determine what information we will look for, and, secondly, what methods we should use to obtain it.

Regarding the first point, we would have learned concepts from our chemistry classes that would directly apply, based on finding information for other reactions. However, in our daily lives, with many other projects and activities, we acquire knowledge about things with approaches similar to what we need for our current Goal.

Let's use the previous cake example in which we are adding raspberries to the recipe. We would first try to remember if we had made this kind of cake with raspberries, or any type of cake with some kind of berries. If we don't have any previous experience that pertains, we would probably then search in the recipe books that we have or on the internet for a recipe or some kind of information about the exact cake that we want to make with raspberries. If we don't find that, we might try generalizing with regard to the cake and the berries, so we look for information about a different kind of cake with raspberries, or the same kind of cake with a different kind of berries, or a different kind of cake with a different kind of berries. With still no success, we might expand our search to a different kind of dessert, maybe a tart, and a type of fruit other than berries. After that, we might contact a friend that likes to bake or even call a bakery. As a last resort, we would probably make our best guess as what to do, try it, and then make revisions to our modified recipe after we taste the product.

As another example, we might be working on our car, and we need to change the engine thermostat, although we have never done this before. We might be able to download a repair manual for the exact make, model, and year of our car that describes and shows how to do this. If it is not available, we search the internet for text descriptions or, preferably, look for videos on the internet that show how to do this for the exact make, model, and year of our car. With no luck, we expand to the same make and model, but a somewhat different year. Next, we try the same make, but a different model. Then, finally, we look for a video that covers changing the thermostat for any car, or a general text-based document about changing thermostats in vehicles. With no luck so far, we would probably consider contacting a friend with auto repair experience or the mechanic at a repair shop.

Notice for these examples that if we don't find information for the exact task that we have, we consider generalizing and expanding our search to cakes or cars that are similar, but somewhat different than what we are working with. We are using analogy to branch out, and we will need to use analogy and imagination to make use of this information if we do find it. We also need to decide what sources to use to find the information, and we must make decisions about the order of trying these sources. So, we are creating, evaluating, and selecting options for our information sources.

But how do we know how to do what I have just discussed in these examples? We weren't born with this ability but acquired it from a large amount of practice observing, doing, making, or repairing things. Growing up, our parents, teachers, and friends explained and showed us what they knew how to do. We learned about gravity, what happens when two bodies collide, motion and acceleration, the cycles of the day and year, weather, the flow of liquids, the force of the wind, etc. We built models, rode and repaired bicycles, did chores, played with chemistry kits, built robots, played sports, repaired cars, learned to play instruments and take care of them, and did experiments in biology, chemistry, and physics classes, etc. As we acquired experiences, we were able to generalize what we were learning and then apply general concepts to new activities. We discovered patterns and causal relationships in many of our observations. And maybe we decided to attend college as a science major, so we took many advanced courses that improved our understanding of the underlying theories of our chosen field and provided practical laboratory experiences.

Through all of these activities, we not only learn a lot about how the world works, but, hopefully, we also learn to generalize the concepts and become increasingly better at using analogy and imagination to adapt to new experiences. With our imagination, we can borrow steps from one series of Events, procedure, or recipe, and integrate them with steps from another series of Events, procedure, or recipe.

We can adapt causal relationships from one situation and apply them to another to allow us to predict the outcome. And we can see that patterns in one setting are similar to the patterns in completely different settings.

An AI with AGI capabilities would need to understand most or all of these same concepts. A humanoid robot could learn these concepts in the same way that a human does. However, depending on the AI software architecture, it may be possible to initially program many of the concepts into the AI to put it in a state that it understands how the world works and is ready to learn new information and make use of it to make new predictions and accomplish new tasks.

As we proceed with the current project, we are applying many of these learnings not only to our current Goal of finding information about our reaction, but also to the original Goal of studying the reaction.

We should first think about this reaction from a general point of view. We know from our chemistry classes that propionic acid is a carboxylic acid and that ethanol is an alcohol. So, we search the internet for reactions of carboxylic acids and alcohols, and we learn that these two types of chemicals do react with each other to form esters by what is called an esterification reaction. The other product is water. Based on the naming convention in the article that we found, the ester product for our specific reaction is ethyl propionate. We can now fill in the blank that represents the "effect" for our cause-and-effect statement and the question mark for our reaction equation:

- If propionic acid and ethanol are reacted together, ethyl propionate and water are formed.

The O-H from the acid and the H from the H-O of the alcohol are removed, forming water, and the remaining parts of these two molecules join together to form what is called an ester link. We also learn two other points from the article. Esterification reactions do not go to completion and the amount of conversion of the reactants is dependent on the temperature and the ratio of the acid to the alcohol. Also, a catalyst is usually required.

Based on our general chemistry knowledge, we would also like to know some of the properties of the four chemicals involved in the reaction, such as molecular weight, boiling point, melting point, density, heat capacity, and heat of vaporization. The internet likely has this information. If not, then specific websites, reference books, and process simulation programs are good resources. For a real project, we would also be researching safety information regarding flammability, reactivity, corrosivity, health effects (breathing, eyes, skin, carcinogen, teratogen, etc.), waste disposal of products, etc. However, to reduce the complexity of this example, we will avoid considering these important topics.

We now know general details about our reaction, but we still do not know what specific conditions are required. Searching on the in-

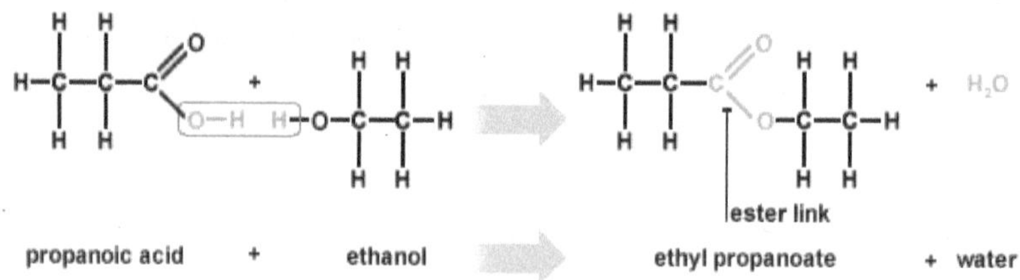

Figure 22. Esterification reaction of propionic acid and ethanol to form ethyl propionate and water.

ternet for our specific reaction (with propionic acid and ethanol) does not yield any results (it actually would, but we are pretending that it does not for this example). We decide that we should consider the use of analogy to obtain approximate details about the conditions. Since we know from the internet article discussed above that carboxylic acids and alcohols undergo esterifications, maybe we can find a paper about the reaction of an acid and an alcohol that are different, but not too different than our compounds.

We find a paper about the reaction of acetic acid with methanol to form methyl acetate (see Figure 23). Both acetic acid and methanol each have one less carbon (and the associated hydrogens) than propionic acid and ethanol. It is reasonable to think that these other reactants, their required reaction temperature, catalyst concentration, and the resulting reaction rate should make a good enough analogy to our system for an initial experiment. After we observe the results from the first experiment, we can make adjustments to the reaction conditions as seem appropriate.

Upon reading the paper, we discover that the reaction with acetic acid and methanol was run at several temperatures and catalyst concentrations. In all cases, equal amounts of acetic acid and methanol were used (to be more technical, the molar ratio was the same, but I don't want to get bogged down with this detail for this example). The reaction rate varied with the conditions, but there were two patterns that were evident. (1) the reaction rate would start out relatively

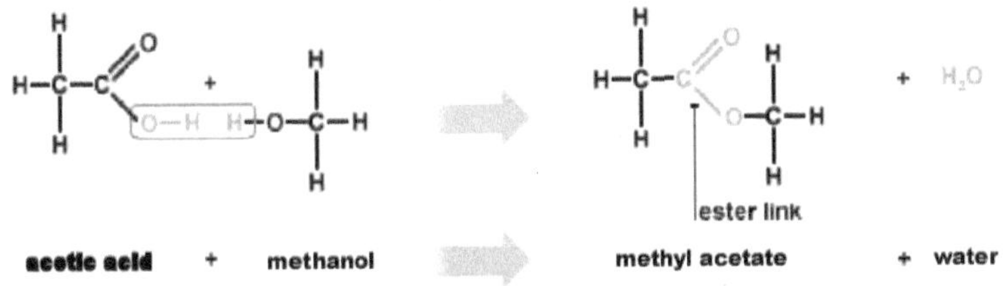

Figure 23. Esterification reaction of acetic acid and methanol to form methyl acetate and water

fast and slow down as the reaction progressed, and (2) the reactions would never go to completion (as was stated in the general internet article about esterification reactions).

We now have two sets of decisions to make: (1) the exact conditions to use and (2) the equipment that we will use to carry out the experiment, which includes the instruments required to determine that we are at the correct conditions and how the reaction is progressing during the experiment.

For each of these decisions, we need to create, evaluate, and select options. We want to use one of the sets of conditions from the paper so that we can compare our results using propionic acid and ethanol with the acetic acid and methanol used in the paper. We do not want to extrapolate any more than necessary for our first attempt and we can use the data from the paper as a check on our experiment. Even though we expect somewhat different results, if they are drastically different from those in the paper, this is an indication that there is likely a problem with our experiment.

Conditions

We know from our previous laboratory experience and the information from the paper that the key conditions that we need to determine are:

- Temperature
- Catalyst concentration
- Reaction time
- Sample schedule
- Liquid volume

The temperatures used in the paper were 40 C, 50 C, and 60 C. We do not have to create the options for the temperature because these have been provided by the report, but we do need to select the best temperature. We want the reaction to proceed relatively quickly, but if it is too fast, we will not get accurate data at the beginning when the reaction is the fastest. We would like to minimize the concentration of catalyst used. Based on our experience with determining reaction rates from our chemistry classes, we determine that the initial reaction rate will be about right with the highest temperature (60 C) and the lowest catalyst concentration (0.1 wt.%). So, we have selected the options for both temperature and catalyst concentration.

From the paper, the reaction time until the composition had stopped changing for the acetic acid and methanol system at 60 C and 0.1 wt.% catalyst was 4 hours. However, since we know that the reaction rate will be different for the propionic acid and ethanol system, although we do not know if it will be faster or slower, we will plan for double that amount of time, or 8 hours. If, after the first experiment, we find that the reaction was too slow, we know from our knowledge of reactions that we have two main options. One of the many cause-and-effect relationships that we are aware of (from our chemistry training) is the very approximate rule-of-thumb that raising the reaction temperature by 10 C doubles the reaction rate. We also know that if the catalyst concentration is doubled, the reaction rate will double. If necessary for the second experiment, we can use these relationships to predict how to change the reaction rate based on data from the first experiment, and we can evaluate these options to determine which will be best.

We will follow the same sample schedule as was described in the paper, which is sampling at 5, 10, 20, 40, 80, 160, and 240 minutes.

The only variable that we have not determined is the liquid volume. We know that the sample will require 1 to 2 ml (milliliters) each, so we want the starting reactor volume to be large enough that the samples do not cause a significant depletion in material (and volume). From previous experience, we think that the reactor volume should be at least 500 ml. However, we also want the reactor volume to be not much more than 500 ml so that we minimize the use and disposal of chemicals. In order to determine the vessel that we will use, we need to determine what is readily available.

We have determined the required conditions, and now we must determine the equipment that we will need and how we will know the concentrations as the reaction progresses. However, to make sure that we do not miss any important details, we want to make a diagram of the equipment (this could also have been done earlier).

Regarding the analyses that we will need, we talk to the analytical group and explain the chemicals that we will be working with and their approximate concentration range. They agree that they can provide analytical support. For an actual project, the analyses would be a project in itself. The correct instruments need to be selected, configured correctly, calibrated, and sample preparation methods developed. But all of this analytical work would be based on many previous projects and training, with relatively routine changes to each part of the process. So, the analytical technician is using analogy and imagination based on previous experience, with some prediction based on causation and patterns.

Equipment

Now we need to determine the equipment required based on the conditions that we want to use for the experiment. However, we also need to determine where and when we will do the experiment. The following list shows the conditions and activities that are the basis for the equipment:

- Liquid volume ≥ 500 ml
- Temperature 60 C
- Catalyst concentration 0.1 wt.%
- Sampling 1-2 ml/sample
- Location
- Time

The main piece of equipment is the vessel, of which we would like to be able to see inside, we will need to be able to add liquids to it, we will need to be able to heat the liquid to 60 C and measure its temperature, and we need to be able to take samples from it while it is at reaction temperature. We also know that the reactor contents should be well-mixed to make sure that the liquid composition is uniform. To collect samples, we have sometimes used a small syringe in previous experiments.

Based on previous experience, Pyrex glass vessels are usually used because of the visual aspect, but also because they can withstand being heated without breaking. An image pops into our mind of a spherical glass vessel, called a round-bottom flask, with an electric heating mantle that covers the bottom half of the flask, and associated controller for the heating mantle. This is a common combination that we have personally used many times. One other point is that we normally fill the flask about ½ full of liquid, so the size of the flask needs to be $\geq$ 1000 ml.

Temperature measurement is routinely done with a thermocouple connected to an instrument that displays the temperature. However, the vessel must have a thermowell which is a narrow glass tube with a closed end that extends almost to the bottom of the vessel. The thermocouple is inserted into the bottom of the thermowell with a small amount of silicone oil for good heat transfer. While we think about this, we are visualizing the equipment that we used for the last experiment: the thermocouple inserted into the thermowell, the plug at the end of the thermocouple cable that plugged into the back of the temperature indicating instrument, and the digital display on the front.

We will need to obtain the propionic acid, ethanol, and catalyst for the experiment, but we need to locate the reactor vessel so that we know the exact volumes that will be needed.

The lab that we will work in must have a fume hood to prevent us from breathing any hazardous vapor. It also needs to have adequate benchtop area for preparing for the experiment and working with the samples.

Regarding time, the experiment will require a full day, and we would like to be able to run it ASAP. However, we plan to run several experiments.

Now that we know the conditions and the type of equipment and lab space required, we need to locate these. We will need to create, evaluate, and select options for each piece of equipment. For the reaction vessel, we have the following selection criteria, which are Attributes and SubEntities:

- Pyrex glass
- ≥1 liter
- Charge port
- Sample port
- Thermowell
- Relatively new in appearance
- Heating mantle with magnetic agitator that fits flask is available

We first locate several 1-liter and 2-liter flasks, but we focus on the 1-liter flasks. One of them is lacking a thermowell. Another does not have a port for taking samples. The third one has all of the necessary ports but has several scratches. One of the 2-liter flasks appears to be relatively new and has all of the ports that we need, and a heating mantle is available.

None of the 1-liter flasks meet all of our criteria, but the 2-liter flask does. We preferred to have a 1-liter flask, but it is more important to satisfy all of the selection criteria, especially those related to safety (no scratches). However, the choice of the 2-liter flask increases the amounts of propionic acid, ethanol, and catalyst that we will need.

We also locate a temperature indicating instrument and thermocouple. We have not used this instrument before, and it is configured somewhat differently than the others that we have used, so we use analogy to determine how to operate it. We locate the jack for the thermocouple on the front instead of the back of the instrument. However, the appearance of the jack itself is almost identical to the one on the instrument that we previously used. We plug in the thermocouple and plug the power cord into the wall. We locate the power button, which toggles right-left instead of up-down from our previous instrument. However, we use previous experience (use of power switches), causation (pressing the button toward ON will supply power to the instrument), and analogy (relate operation of previous instrument to current one even though there are differences) to understand that we should toggle the switch in the direction of the word "ON." Doing so, the display lights up and reads 21.5 C.

We locate several heating mantles for 2-liter flasks, but only one of them has a built-in magnetic agitator, so we select that one. Upon testing, we find that it works as expected.

Sample bottles and a sampling syringe are available. We determine how much of the reactants and catalyst are needed and order them.

MultEvent and SubEvent Options for Procedure

Now we know the conditions and have the equipment located. We need a procedure, or MultEvent, to allow us to progress from the initial state to the Goal state. There are several possible options for procedure development:

1. Someone provides the procedure
    a. Low level analogy, causation, and imagination required
2. Previous experience with a similar experiment
    a. More analogy, causation, and imagination required
3. Taking parts of other procedures and concatenating them
    a. Higher level of analogy, causation, and imagination required

For the first option, we obtain the procedure from someone or from a written document, and we just need to follow it. However, this option will still require a fair amount of cognition because the procedure probably will not exactly describe our equipment and our situation. If the procedure states to pour the reactants into the flask, we still need to know physically how to set a beaker on a balance, tare the balance, reach for one of the reactant bottles, grasp it, unscrew the cap with the other hand, lift the bottle, tilt it over the beaker, pour the reactant until the balance reads the correct weight, etc. There will be many other small details that we will need to know how to do that will not be in the procedure, such as placing the flask into the heating mantle, how to turn on and set the heat input for the heating mantle, how to place the thermocouple in the thermowell, how to turn on the temperature indicating instrument, etc.

For each of these actions from the procedure, we will do something very similar, but with different equipment, different bottles, different instruments, different heating mantles, etc. We will need to use analogy and imagination to relate our previous experiences to the new equipment. And we are also using cause-and-effect relationships:

- When we turn on the heating mantle, it will warm up and heat up the liquid
- When we turn on the agitator, it will rotate
- When we pull the syringe plunger, liquid will be pulled into the syringe
- When we combine the reactants and catalyst and heat it up, the reactants will react
- If the temperature increases, the reaction will go faster
- The reaction will slow down as it progresses
- The reaction will stop before it is complete

For the second procedure option, we will have similar activities as just described, but we will need to make some modifications to the procedure that we used previously. We will likely have different chemicals, quantities, temperatures, and times, requiring these details of the procedure to be modified, although the procedure steps may not change otherwise.

Alternatively, let's assume that we have not run an experiment just like this one before and that we need to create the procedure, which would represent the third option. We might have a procedure that can be used as a starting point and that we just need to modify some parts, or we might take the parts of several smaller procedures and concatenate them together.

Let's assume that we have the last situation, so we need to create the procedure from scratch, but we have parts of other procedures that we can make use of. The procedure is the overall MultEvent, with is a series of SubEvents, each of which is hierarchical in that they can each be divided into a series of lower-level SubEvents.

To create the procedure, we need to review what are the initial and Goal states. The initial state represents the chemicals being in their original bottles, the equipment laying on the table unassembled and not plugged in, and the temperature of everything being at room temperature. The Goal state represents the equipment being assembled and plugged in, the chemicals being in the flask, the chemicals being at the reaction temperature for long enough that the reaction has proceeded and stopped, samples have been collected at specific times and analyzed, the resulting process and analytical data has been analyzed and the results summarized.

Based on our experience with other experiments, with any cooking that we have done, and projects in general involving equipment, we know that we should have the equipment assembled before adding the chemicals. We know that the chemicals must be brought to 60 C for the reaction to proceed as desired, but that the chemicals need to be in the flask so that the heating mantle can provide heat, so the chemicals will need to be added to the flask. Then the heating mantle will need to become hot in order to heat up the chemicals, and we will need to only add enough heat to get to 60 C and no more. To know how the composition of the chemicals is changing, we will need to take samples at specific times and have them analyzed. And we will need to continue until the composition is not changing anymore.

So, the MultEvent, or overall procedure, can be divided into several SubEvents, or parts:

1. Assemble equipment
2. Add chemicals to flask, except for the catalyst
3. Heat flask to 60 C
4. Add the catalyst to the flask and start a timer
5. Sample reaction liquid at 5, 10, 20, 40, 80, 160, and 240 minutes
    a. Maybe longer if the reaction has not stopped by 240 minutes
6. Have samples analyzed

Each of these SubEvents would likely be based on something that we have experience with in some form that is similar enough that we can make an analogy to the current SubEvent (if a given SubEvent is completely new, then we would need some type of training, but we will assume for this example that we have had some hands-on experience with each of the parts). But the experience that we have with each SubEvent may be from several different unrelated projects. So, we are essentially implementing Option 3 from the beginning of this subsection, in which we are assembling the overall procedure (MultEvent) from parts of other procedures.

Let's consider, for a moment, other options for the order of these five SubEvents, even though our ample experience should almost automatically select the order shown above. We might consider adding the chemicals to the flask before the equipment is assembled. We have now created two options for the order of the first two steps. The next step would be to evaluate the two options. Some of the considerations for the second options would be:

1. The flask would need to be supported to pour the chemicals into it (instead of sitting in the heating mantle for support).
2. This option would allow the flask to be set on a balance so that the chemicals could be weighed as they are poured into the flask, which would eliminate some transfers (a moderately positive point)
3. It might take some time to assemble all of the equipment, including plugging in the temperature indicating instrument, connecting the thermocouple, and checking that the instrument reads correctly. The heating mantle would need to be plugged into its controller and tested. If this takes very long then the chemicals may have started reacting before the beginning of the experiment, even before catalyst is added (a significantly negative point).

Based on just these points, with the SubEvents grouped as they are, the last point would tend to push the selection toward the first option. However, it might also cause us to repeat the original option creation step for the overall procedure, and to consider breaking up the first and second SubEvents, "Assemble Equipment" and "Add chemicals to flask, except for the catalyst" as follows:

- 1a. Assemble all equipment except for the insertion of the flask into the heating mantle and the insertion of the thermocouple into the thermowell
- 1b. Place the supported flask on the balance
- 2a. Weigh the chemicals as they go into the flask, except do not add the catalyst

2b. Place the flask in the heating mantle and place the thermocouple in the thermowell
3. Heat the Flask to 60 C
4. Add the catalyst to the flask
5. Sample reaction liquid at 5, 10, 20, 40, 80, 160, and 240 minutes
    i. Maybe longer if the reaction has not stopped by 240 minutes
6. Have samples analyzed

After we have decided on the overall MultEvent, we would then need to create the more detailed, secondary lower level (hierarchically) SubEvents. And we would need to create possible options for these SubEvents. The following shows one of the option sets:

1a. Assemble all equipment except for the insertion of the flask into the heating mantle and the insertion of the thermocouple into the thermowell
    i. Place the heating mantle on the table in the fume hood
    ii. Place the heating mantle controller next to the heating mantle
    iii. Plug the heating mantle into its controller
    iv. Plug the controller into the 110 V outlet
    v. Place the temperature indicating instrument on the table near the heating mantle in a location where the digital display is visible through the hood doors
    vi. Plug the thermocouple into the temperature indicating instrument
    vii. Plug the temperature indicating instrument into the 110 V outlet
    viii. Turn on the temperature indicating instrument
    ix. Observe the digital temperature display
1b. Place the supported flask on the balance

Other options could be created for the 1a SubEvents, which might be based on a different order for some SubEvents, such as additional checks on the equipment, or for other reasons. The evaluation of each of these options would be based on time, efficiency, safety, or other SubEvent Attributes, which would be used to select the best option for the 1a SubEvents.

The rest of the 6 high-level SubEvents would be divided similarly into secondary lower-level SubEvents that would go through the option creation, evaluation, and selection process.

Most, or all, of the secondary SubEvents would be split further into tertiary SubEvents, in which "Plug the controller into the 110 V outlet" might be divided into:

- Reach for the plug
- Grasp the plug
- Move the plug toward the outlet
- Push the plug into the outlet

At this level, some of the SubEvents are carried out almost subconsciously.

After the experiment was conducted, the results would be evaluated relative to the reaction rate expected. If it deviated significantly, then the experiment should be repeated with a different set of conditions. Either the temperature or the catalyst concentration could be changed, so a decision would need to be made as to which one. If temperature was chosen, then the cause-and-effect relationship of "the reaction rate doubles every 10 degrees" could be used to help with the prediction of how to change the temperature.

These examples, although somewhat oversimplified, provide an overview of the process that a human goes through in order to accomplish a goal to build, make, or learn about something. There are numerous other types of Goals, which might vary greatly in degree of complexity, but would require the same type of thought processes covered in this chapter.

The goal of this chapter was not to consider how humans go about accomplishing things for its own sake, but to review the main points involved with these activities to serve as a reference for an AI that can perform at a level representing AGI. For an AI to perform at the same level as a human, it would need to be able think through the problem and its physical aspects and then implement steps to accomplish the Goal in a similar manner as described by the examples presented.

Now I would like to summarize the capabilities that an AI with AGI would need in order to carry out the examples described in this chapter. If the AI were to be able to physically and mentally accomplish the tasks from these examples, it would likely be in the form of a humanoid robot. If it only worked at a specific location or station, it may only need to resemble the top half of a human, or if it only provided instructions for an embodied AI or human and then analyzed the results from the activities, this could be accomplished with something like a desktop computer, or even a phone. But I would like to present the summary based on the concept of a humanoid robot since this would be the most capable. The AI would need (at least) the following capabilities:

- Sight, hearing, touch (smell may also be required; taste would be optional except for occupations related to food)
- Agile control of its body to stand, walk, reach, grasp, lift, climb stairs and ladders, pour from and fill up containers, hold and manipulate delicate Entities, etc.
- Expansive knowledge of cause-and-effect relationships, based on experience, observation, and reading
- Large knowledge base of recipes and procedures for doing and making things
- Extensive experience designing, assembling, making, and operating equipment and devices
- Able to detect patterns, and interpolate and extrapolate based on them
- Extensive abilities to solve problems using various forms of math
- Able to use physical and software tools
- Able to learn new cause-and-effect relationships and procedures by observation (from other robots, people, videos) and reading
- Use analogy to relate experiences, abilities, and existing knowledge to new situations
- Use imagination to mentally modify, combine, and manipulate existing knowledge in a way that allows for the mental creation of new procedures and Entities that can be used to achieve a Goal
- Able to use cause-and-effect relationships, patterns, analogy, math, and imagination to predict outcomes
- Able to convert needs from itself and those of outside requests into Goals that will satisfy these needs

- Can create, evaluate, and select options for action plans to achieve Goals
- Able to prioritize competing Goals
- Can evaluate the results of carrying out tasks of a Goal based on the actual results compared with the Goal

13

Consciousness and Self-Awareness

"Any Entity that can perceive the present, remember the past, and contemplate and predict the future is a conscious and intelligent being. And if that Entity also knows that itself is the one doing the perceiving, remembering, contemplating, and predicting, it is also self-aware."

- Glenn Graham

Consciousness

I think that it is entirely possible for an AI to be conscious in a manner not too different from that of a human, based on the discussions presented in the previous chapters. However, there have been and are numerous theories on what consciousness is, so we need to sort through these ideas and think about what it actually means for a human to be conscious from a functional point of view. In other words, I would like to consider what we humans actually think about on a daily, hourly, minute-by-minute basis as the manifestation of our consciousness. And if we can create an AI that is able to think in a similar manner, then it should also be conscious.

In this chapter, I will first give a brief overview of the concept of consciousness as has been studied and written about by past authors. Then I want to discuss what it means for a human to be conscious and provide some thoughts about how this relates to potential consciousness in AI Entities.

Background

Consciousness is defined in the Merriam-Webster English Dictionary as [47]:

1a the quality or state of being aware, especially of something within oneself

2a the state or fact of being conscious of an external object, state, or fact

According to Victor Caston, Aristotle (384-322 BC) had a concept of consciousness that was similar to perception [48], and he believed that "consciousness is both intrinsic and higher order, due to its reflexive nature."

John Locke (1632-1704) thought that the mind was a "blank slate" (tabula rasa) when we are born and that consciousness develops through experiences [49] [50] [51]. He stated that, "Consciousness is the perception of what passes in a Man's own mind." He also considered memory to be an important part of consciousness because linking past and present experiences forms the self.

Somewhat like Aristotle, David Hume (1711-1776) thought that the self was essentially a bundle of perceptions [52] [53] [54]. He argued that consciousness is the combination of constantly changing perceptions, sensations, and thoughts, instead of a unified, enduring identity. A person's concept of themselves is the summation of all of one's individual experiences reviewed by the mind in rapid succession.

Edmund Hesserl (1859-1938) inaugurated the school of thought known as phenomenology, which "attempts to describe the universal features of consciousness while avoiding assumptions about the external world [55] [56]." It attempts to understand the meaning of the lived experience and is sometimes referred to as the "science of experience." Husserl thought that consciousness is the awareness of things other than the self, whether these things are physical or imagined [57].

Existentialism is an intellectual tradition that looks at the struggles of individuals trying to lead an authentic life despite the "apparent absurdity or incomprehensibility of existence [58]." Jean-Paul Sartre (1905-1980) was one of its key proponents and believed that an individual's actual life is what constitutes their "true essence" rather the essence attributed to them by others [59] [60]. Using their own consciousness, people create their own values and determine meaning in their lives.

Still today, there are many theories of consciousness [61]. However, as presented by Ralph Lewis, there are four theories that are some of the most significant, and I have them quoted here from Reference [61]:

- **Higher-order theories (HOTs):** These theories propose that thoughts become conscious when basic perceptions ("lower-order" representations) become re-represented as higher-order representations at higher levels of the brain, specifically in the prefrontal cortex.

- **Global workspace theories (GWTs):** According to these theories, perceptions, thoughts, emotions, etc. become conscious when they gain access to a "workspace," or to apply another analogy, when they make it into the "spotlight"—as if the mind were a theater where conscious thought is the activity in the spotlight on the stage at a given moment.
- **Integrated information theory (IIT):** This theory is about how parts of a brain interact to create a unified experience. Consciousness is theorized to be related to how much information is integrated among the different parts of the brain. IIT tries to mathematically measure this. The more information that is connected and integrated, the more conscious the system is thought to be. The theory suggests that not just brains but any complex system with the right level of interconnectedness and integration of information could exhibit consciousness, with simpler systems manifesting only an insignificant modicum of consciousness. IIT attempts to explain the *quality* of conscious experience, suggesting that the unique pattern of relationships between the elements of the system is what defines a particular conscious experience.
- **Re-entry and predictive processing theories:** The core claim of re-entry theory and predictive processing is that conscious mental states are associated with top-down signaling. Top-down signaling refers to the process by which higher-level brain regions send information, expectations, or context to lower-level brain regions. This communication helps shape how the brain perceives and interprets bottom-up information received via its sense organs from the world around it. (Predictive processing theories are not specifically theories of consciousness but are more general accounts of brain function that can be used to formulate ideas about the properties of consciousness.)

What we think about

As discussed above, humans have been intrigued with how and what they think about probably for as long as they have been thinking. Many theories have been developed in which human thought processes have been studied from numerous angles. But I believe that consciousness for humans is really just the combination of all of the things that we think about on a daily or weekly basis, because whatever we are thinking about is essentially the definition of consciousness. This chapter pulls together most of the concepts covered throughout the book.

I would like to look at the concept of consciousness from a practical viewpoint based on what we as humans experience with our brains. Our level of consciousness is the highest known on Earth and possibly the highest in the universe. The question remains as to whether we would want AIs to be conscious, but if we do, then basing their consciousness on that of humans is probably as good of a starting point as any.

So, let's review the previous chapters and reverse-engineer our own minds a little bit and try to understand consciousness from that perspective. For starters, everything that we think about takes place in the past, present, or future. In the present, we have the input from our five senses and emotions that we feel because of what we experience. If you are by a lake in the Rocky Mountains in the summer, the width and height of the mountains fill a large fraction of your visual field. The shimmering blue lake glistens in the sunlight, and you see hundreds or thousands of tall green trees covering much of the landscape. The motion of a variety of birds, small mammals, and insects, around and above, pleasantly distracts you. You may even see deer, elk, or a moose off in the distance (hopefully no bears). You hear the sounds made by the animals, and if there is a breeze, the rustling of the tree limbs. If it's early in the day, you probably feel a chill in the air, and the air has a subtle, but fresh, smell. And when you pull out the snack that you brought, it likely tastes just a little bit better because of everything else just described.

At the same time, you are forming memories of what you are experiencing. And you are likely feeling emotions from what is happening around you. You might feel happy from the sensations of the present just to be in such a beautiful place. Or a little upset at yourself that you forgot to bring a warm enough jacket, if you are cold.

In addition to the emotions you feel, you likely also have desires, some of which will stem from those emotions. With the current example, the happy feeling of being in the mountains might give you the desire to return there again next year. Or you might notice that the tread on your hiking boots is getting worn out, so you desire to buy a new pair.

We also spend a lot of time remembering the past. For the current example, you are almost certainly recalling memories of the last time you were here, or of a similar place, or of the photos that you saw online of this location, or someplace similar. But the memories might also be more indirectly related. For example, you might think about where you were at this time last year, or the time you were here with one of your siblings, or the moose you saw off in the distance about a mile back down the trail, or even the list of what you need to do when you get home.

And there are often emotions associated with our memories. While remembering your sibling, you might feel sad that he/she isn't there with you (or glad if he/she is a complainer). Or you might have a feeling of awe thinking about the moose you saw. Or despair remembering your to-do list and everything you need to do when you get home.

Of course, we think about the future a lot. The desire that you had to return to the same location next year will likely make you start planning the trip. Or you might already have a trip planned for next month to a country in Europe that you have never visited, and you are anticipating what you might see and experience. And you'll probably start thinking about where to get the new boots that you have a desire to purchase. Or you might be thinking ahead about the dentist appointment next week to extract a tooth or [[figuring out how to do the next experiment at work, or creating a to-do list in your head, or thinking about how to better organize your garage. Each of these thoughts about the future were initiated by a desire and could also have emotions associated with them.

Much of what we experience and think about relates to the Self, and, therefore, self-awareness. While we are experiencing, remembering, and planning with regard to everything around us, we are also thinking about how our Self is interacting with the world. As we experience the mountains, lake, trees, and animals, we are aware of the blister on our left heal, the chilly feeling from the breeze, where we are walking to avoid tripping, how the backpack feels, how we feel about the new co-worker at our job, or some disturbing news that we learned about yesterday, and we are also thinking about what we are thinking about.

We tend to combine the past, present, and future in different ways. Consider a vacation trip, as an example. We decide where we would like to go. We search the internet for information and photos. We plan the flights, hotel room, and the activities for when we are there. Then we go on the trip and moment-by-moment, our five senses are bombarded with stimuli. After the trip, we think back and remember the experiences we had while on the trip and feel once again the associated emotions.

Or you might be building something, maybe a woodworking project. Depending upon the part of the project, you might be thinking about the past, present, and future almost simultaneously. If you are trying to assemble two wood pieces, you might be remembering a similar step on another project, planning specifically how you are going to do the assembly, attempt to make the assembly, but find it is not going together as well as expected, so you are trying to remember in more detail what you did before, thinking about how that knowledge relates to the current situation, and physically trying to get the pieces together.

In some manner, we are learning from each of our experiences. For whatever we planned earlier and carried out, we are probably evaluating if things went according to plan. If it did, then we can repeat it again later to hopefully get a similar result. If things went awry, then we know to try not to repeat the same Events, and we can think about what changes we need to make for the next time.

These are the kinds of things that we think about on a daily basis. However, I have only presented what we think about but not how these thoughts are related, how we focus on specific thoughts, and how these points are related to consciousness.

Spotlight of Attention

Our brains are able to store a tremendous amount of information, with estimates as high as 2.5 petabytes (2.5 million gigabytes) [62], which is about 120 years of 1080p movies. And the data transfer rate from the retina to the brain is about 10 Mbps, which is about the same speed as Ethernet [63]. However, the data processing rate for information withing the brain is only about 10 bits per second [64]. And we all know that it is difficult to focus on more than one thing at a time, if these things are very complex.

The term "Spotlight of Attention" refers to how we focus our attention on an important set of details (like a spotlight beam), while ignoring surrounding, or unimportant, information. So, we can only think about a small fraction of the information arriving through our senses or potentially being retrieved from memory. And our brain has to choose what set of information to focus on. In addition, whatever we focus on also get stored in our memories

Let's think about how and why we focus our attention. The main goal here is to understand what our brain is doing, not necessarily exactly how it does it. The ultimate goal is to be able to develop an AI that can think in a way that mimics how a human brain processes information, even if it is not actually processing information in exactly the same way. It's a little bit like the bird-airplane wing comparison; the bird's wings have feathers and flap, while the airplane's wings are metal and are fixed on the plane, but they both can fly.

Most of the concepts that will be discussed were covered throughout the book, so this discussion will tend to pull together and summarize them. In addition, I will refer to the Spotlight of Attention as simply the Spotlight.

Our Spotlight acts like a control room that all information flows in and out of. At any given time, our Spotlight is bombarded with information from our senses and our memory. We also have a mental To-Do list of Goals, covering everything from "I want to look at the gorgeous view," to "I need to eat lunch," to "I need to write the proposal," to "I want to learn a new language," which must be prioritized so that the most important Goal is focused on first. In reality, we normally have numerous Goals that we are trying to accomplish at the same time, and we are in different stages of completion for each. However, our mind still must determine which of these has the highest priority and the steps for each of the Goals must be prioritized relative to the steps of the other Goals. But for the following example, to keep things simple, one main Goal will be prioritized over the others.

Consciousness and Self-Awareness Example

At any given time, we normally have hundreds of Goals that cover a wide range of importance levels and time frames, including everything from "I want to eat that donut now" to "I want to eventually become CEO of the corporation." But to keep the focus on consciousness and self-awareness, I would like to consider an example in which we only have the following four Goals:

- Experience the scenery
- Answer a question
- Build a piece of furniture
- Plan a trip

Since we are by the lake in the mountains (see the example under the **Consciousness** – *What we think about* section), and the other Goals can wait, we will focus on the first Goal. Our mind is being bombarded with input from all 5 senses, but only some of this information is being processed. We see the trees and part of the lake and mountains that are directly in front of us, but only what we focus on. We hear the sounds of birds and flying insects, the chatter of squirrels and chipmunks, and the sound of the wind rustling the leaves and branches, but we will focus on some of the sounds and miss others (similar to being in a crowded room of people talking and focusing on one conversation). We feel the cool breeze, the minor pain of the blister on our left foot, and the weight of our backpack, our clothes against our skin, and the minute pressure of our glasses sitting on our nose, but we probably don't notice all of these sensations all of the time; only when we focus on them. And maybe we can smell the pine needles and taste the inside of our mouth, if we pay attention to these sensations.

Some of these sensory inputs are processed in parallel and some in series. We might see a hawk fly overhead and hear its cry at the same time. But if there are several hawks, we may focus on one or two, then a different one, and back again.

Whatever is in our Spotlight will be remembered and will be processed to identify it. The hawk that we are focused on will be compared with memories we have stored, which will be 2D and 3D images and videos. Our mind will be able to determine that it is an animal that flies, and that it is a bird. If we have detailed knowledge in our memory banks of birds, we may be able to identify the bird as a hawk, and possibly even what type of hawk. And we will remember that we made this identification and what the bird was doing.

When we focus on the trees, our mind will compare what we see with our memories in a similar manner to the bird. The same for the mountains, the lake, the insects, the other animals, etc. Note that we are using analogy to make these identifications, in that none of Entities will look or be oriented exactly the same as those in our memories.

Similar mental activities will occur for the sensory input from the other four senses. For example, the call from the hawk will evoke memories of other bird calls, which, if they had previously been identified as being made by a hawk, could be mentally used to identify the current sound as being made by a hawk. Here again, analogy will probably be used to make the comparison.

For each of the Entities and their Actions that we focus on, our minds will create Events which are stored in memories which add to the existing memories that were used to identify the Entities and their Actions. However, some of the existing memories may be revised based on the new information.

Emotions are likely already associated with the memories that are retrieved, but new emotions may be evoked for both the old memories and the new ones created by the current observation. This is true especially for visual input but would also apply to the other senses.

For the memories that are retrieved as part of the identification process, there are many associations with other memories, which tends to cause a "stream of consciousness" to occur. For example, viewing the lake would cause memories of lakes to be retrieved, first to identify that the Entity in front of us is a "lake." But one of the memories might be for a lake you visited last summer with Bill and Joe, and you remember that Bill got married a couple months after that, and he and his wife moved two months later, and that his wife had gone to the same university as another friend of yours… The "stream of consciousness" could go on for as long as you allow it, but you can force your mind to redirect and focus on a specific one of the memories or on what you are currently looking at.

In addition to all of the sensory, memory, and emotional input to our brains, we also will have desires that contend for a spot in the Spotlight. This could be hunger, thirst, an itch on the back that you can't reach, a desire to return to this location sometime in the future, a desire to purchase a new set of boots that won't give you a blister, or a desire to climb the hill on the other side of the lake. There could even be a long-term desire that sneaks into your thoughts, such as wanting to get serious about moving to a house with more property, or a desire to change jobs to increase your salary.

Let's assume that the desire to climb the hill on the other side of the lake outweighs the other desires and rises to the top of the Goal list. Your mind uses existing memories of similar settings and climbing experience for hills, along with causation, analogy, and imagination to create options to get to the other side of the lake and climb the hill. This includes which side of the lake to go around and which side of the hill to climb. The path options are evaluated to determine the best path, based on criteria such as distance, time, steepness, openness of the terrain, etc., and then the best path is selected.

All of the while, your brain is creating chronological memories of everything that you have thought about and done. As discussed in a previous chapter, there is one type of memory (real-time personal environs) that tracks the input from the five senses of the surroundings and associated Entities and their Actions, and tracks the recall of related memories (for Entity identification) and derivative memories (from stream of consciousness).

So far, the discussion has covered facets of consciousness with some aspects of self-awareness. But there is another type of memory (real-time self-based) that does the same thing for things specific to the Self and, therefore, self-awareness. Where did your body go, what did it feel, what did you see and hear about yourself, and what did you think about, remember, and plan, and your self-evaluation of previous predictions, decisions, and Actions. The real-time self-based memory allows you to think about what you thought about previously.

AI → AGI or ASI

As mentioned above, this is a simplified version of all that goes on in our mind but covers many of the main concepts. If an AI is able to mimic these same features, it should be capable of achieving AGI or ASI (Artificial Super-Intelligence), and it should be conscious and self-aware. Because of the following advantages that a computer-based system would have over a biological one, once an AI is able to think in the same manner as a human, it will simultaneously have achieved ASI for the following reasons:

- Much faster processing time
- Ability to have many more than five senses (radio, microwave, ultraviolet, ultrasonic, magnetic, etc.)
- Unlimited memory
- Higher resolution for memories
- Almost unlimited associations between memories
- The possibility of multiple, or unlimited, Spotlights of Attention (compared with just one for a human brain)
- Enhanced logic
    - Unlimited causation relationships, greater pattern recognition, more advanced analogy recognition, greater imagination capabilities, greater math capabilities

In addition to these points, there are several other aspects of an AI that would assist with it achieving AGI or ASI, especially for a humanoid robot. One of these aspects is the sense of touch, including pressure and pain sensation, and even better with hot and cold sensation. This could range from sensations in the fingers and hands to sensations for the entire body of the robot, approaching that of a human. This would not only help the robot to interact with its surroundings, but it should also enhance the robot's sense of Self and self-awareness. This is because cause-and-effect relationships would likely be developed between the appendage movements, initiated by the AI of the robot, and all of the touch/pain/hot/cold sensors throughout the AI's body. For example, if the AI-robot was standing next to a wall and it reached toward the wall until it touched it, felt the temperature of the wall, and could not move the wall even with significant force pressing against the wall, the AI-robot would be making associations between the AI-robot Self and its surroundings. As it continues to develop these types of causal relationships and other associations, based on interactions between the AI-robot Self and its surroundings, it will be reinforcing its concept of Self and its self-awareness.

Emotions would not be necessary for many applications of an AI but would be very beneficial for any interactions with humans. In addition, emotions in an AI should help it understand humans in the same way that humans understand other humans, which would be a positive aspect for achieving AGI. I don't think that we want or need an AI to experience all of the same emotions that a human does, but we would like an AI to understand the full range of human emotions and what causes them. For example, we don't want an AI to experience hate or jealousy, but we would like for the AI to understand these emotions in humans and what causes them to occur.

As was discussed in a previous chapter, human emotions serve to encourage a person to do things that are good for themselves and avoid things that are bad for themselves. Emotions are innate characteristics which could be programmed into the AI. Assuming that the AI's purpose is to help humans (hopefully this is always the case in some form), when the AI is able to help someone, an electronic equivalent of the feeling of happiness could be associated with the corresponding MultEvent performed by the AI. That would encourage the AI to repeat that activity, or something similar, again in the future. Likewise, if the task that the AI performs for a person is not helpful, or even harmful, the AI would sense the electronic equivalent of sadness, and it could conduct a self-evaluation to determine what went wrong and attempt to make changes to the MultEvent so that next time its Actions will be helpful instead of harmful. The AI will gradually develop many causal relationships between its own Actions and the emotions that it "feels," which will be associated with the Actions of the AI Self. As with the sense of touch described above, these associations will have the effect of enhancing the AI's concept of Self and its self-awareness.

Some of the desires that an AI-robot could have would likely be much different than that of a human. For example, instead of the desire to find and eat food, a typical desire for a robot would likely be to find a charging port when its battery is low. Or if it had a malfunctioning part, it might have the desire to find the correct replacement part and repair itself, or to go to a location where it can be repaired. But, just as for emotions, an AI should be able to understand the desires that humans have and what the driver for them is. Another desire that an AI should have is the desire to help humans in certain ways. Many of these desires would be innate aspects which would likely be initially programmed into the AI-robot. The correct desires would make the AI more human-like and likely more prone to achieve AGI.

Several authors have expressed concerns that an AI with AGI or ASI might take over the world and even end human civilization. I believe this scenario is possible if world conquest is programmed into the AI. Or the AI could be programmed to not allow itself to be turned off, at any price. In this case, an AI might decide that it needs to kill anyone that wants to deactivate it (think HAL9000). However, if the AI is programmed with desires something like those associated with Isaac Asimov's Three Laws of Robotics [65], there is no reason that the AI would be a danger to society or be anything but helpful to humanity.

The desires that drive some humans to overtake the world are based on relatively simple concepts. We humans want to feel comfortable, have a nice place to live, and to be important. However, some humans take these desires to extremes and strive to achieve them at an excessive level at any cost. If these types of innate desires are not present in an AI, then there should be no danger of the AI wanting to be king of the world or to harm any humans trying to achieve some Goal.

I believe that consciousness and self-awareness can be summarized by the quote at the beginning of this chapter:

> *"Any Entity that can perceive the present, remember the past, and contemplate and predict the future is a conscious and intelligent being. And if that Entity knows that itself is the*

one doing the perceiving, remembering, contemplating, and predicting, it is also self-aware."

However, I believe that the following manifestations, as previously discussed, are especially important for self-awareness:

- Spotlight of Attention
- Real-time Self-based Memory
- Sense of Touch/Pain/Hot/Cold
- Emotions

Therefore, if an AI can effectively experience these same aspects of consciousness and self-awareness as a human, then I would argue that the AI is also conscious and self-aware, although it could be somewhat more or somewhat less conscious and self-aware than a human. However, if the AI has achieved AGI or greater, then because of the enhancements described above (increased speed, memory, causal relationships, etc.), it is likely that the AI could be even more conscious and self-aware than a human.

14

Final Thoughts

"Everything's destiny is to change, to be transformed, to perish. So that new things can be born."

- Marcus Aurelius

The goal of this book has been to present ideas that could, hopefully, lead to the creation of AI with AGI or ASI. The best example that we have of general intelligence, currently, is the human brain. However, if the main cognitive abilities of the human brain can be copied in an AI, then it should have similar cognitive functionality. I believe that it is necessary to mimic the cognitive functions performed by the human brain, but not necessarily by the exact same mechanism, complete with firing neurons. And, if we also copy the functions of the human brain that give rise to consciousness and self-awareness, then the AI on the receiving end should also be conscious and self-aware.

I think that probably the most important aspect for an AI to achieve AGI would be that it perceives the world as 3D Entities and Actions and, therefore, 3D Events. This is sometime referred to a "world model." This conception is critical for fully understanding the Entities themselves and their Attributes, e.g., the size, dimensions, volume, area, length, or weight, and for understanding the motions of Entities. It is also important for the relationship of an Entity's SubEntities, e.g., the positions of the SubEntities relative to each other and how they are attached to the Entity. And for physical Group relationships, such as the proximity of Entities relative to each other (e.g., a flock of birds); or machinery parts fitting together and interacting perfectly; or kinetic effects, such as a bullet hitting its target, an airplane landing, or ocean wave crashing against rocks. It would be difficult, or impossible, for an AI to fully understand these Entities, their Actions, and their interactions without perceiving them as 3D objects.

I also believe that an AI with AGI would need to have the input of at least some of the five senses of human beings to allow the AI to be able to perceive the world in the same manner as a human to support the AI's understanding of the 3D world. Depending on the application of an AI, it would not necessarily need to have all five of the human senses. Possibly vision, hearing, and touch (including texture, vibration, pressure, pain, and temperature) would be sufficient for most situations. Vision and hearing should cover at least the same frequency range as for a human, but it could be beneficial if infrared and ultraviolet, and infrasound and ultrasound, were detectable. As an AI with AGI, I think that it should at least comprehend that a human also has senses of smell and taste, and better yet, the AI would have those senses also. There are many tasks and jobs in which the sense of smell is important (chemical, fragrance, and food industries), and the sense of taste would be critical for jobs related to food preparation. Other types of sensory input could also be valuable, such as radiation, radio communication signals, and GPS signals, but the AI could also use electronic devices in the same way that humans do, to detect these signals.

An AI with AGI needs to have significant capabilities with the use of logic, including analogy, causation, pattern recognition, and math, as well as highly developed imagination abilities to make sense of the world. Causation, pattern recognition, and math are the key tools that allow predictions to be made. Analogy enables the application of existing knowledge, in the form of memories, causation relationships, and patterns to new situations with their inevitable differences. And imagination allows for the mental modification of Entities, Group relationships, procedures, cause-and-effect relationships, and patterns that can be implemented in the real world to create things, make improvements in existing Entities and procedures, and avoid undesirable outcomes.

If an AI is able to mimic the same cognitive functions of a human that give rise to consciousness and self-awareness, the AI should also be conscious and self-aware. In the chapter on Consciousness and Self-Awareness, I discussed the various types of thoughts that occupy a person's mind on a regular basis, and these thoughts represent what it means to be conscious. These thoughts involve experiencing the present, remembering the past, and thinking about and predicting the future. More specifically, we focus our attention on things that are important to us, including our surroundings, things that affect us physically, mentally, and emotionally, as well as what we are thinking about at the moment. We store chronological memories of what we are experiencing and thinking about. Our sense of touch, including pressure, pain, and temperature provide a physical connection between our minds and our surroundings, whereas our emotions help to make mental connections (pain for an AI could be represented by a subroutine that tells the AI to avoid or stop whatever is causing the pseudo-pain sensation). All of these types of thoughts and mental connections in humans represent what it is for a human to be both conscious and self-aware. Therefore, if an AI can mimic these same mental functions, it too should be conscious and self-aware.

An AI should have very expansive capabilities with regard to processing speed, memory size, and resolution of sensory inputs. It should also be able to have nearly unlimited associations between memories, and enhanced logic due to nearly unlimited causal relationships, greater pattern recognition, more advanced analogy recognition, and greater imagination and math capabilities. Therefore, if an AI is able to match the cognitive abilities of a human brain well enough that it achieves AGI, it will immediately have also achieved ASI due to its significant advantages in mental processing speed, memory size, and resolution in all aspects.

I do not believe that we have to worry that an AI with AGI is highly destined to become evil and wreak havoc on the human race. Humans have inherent needs for comfort, pleasure, and a feeling of importance that can become excessive and lead to unchecked desires that can cause people to commit crimes and murder to satisfy their exaggerated needs. AIs do not innately need food, water, warmth, stylish lifestyles, status, or power. So, if these needs and desires are not programmed into them, they have no driving force to become overbearing tyrants or organized crime bosses. However, there are two main ways that these negative outcomes could manifest themselves in an AI. One would be that these desires are programmed in to support the evil purposes of their human overlords. The other would be that an AI with significant political power comes to believe that the best way to provide benevolent care for its subservient humans is to become an overbearing tyrant. Extensive efforts to avoid these negative consequences should be made and regulations should be enacted to ensure that evil or careless desires are not implemented in advanced AI systems.

Currently, there is a significant concern about the loss of jobs to AI, and the main expression of this apprehension is by high-paying white-collar knowledge workers, such as programmers, authors, and financial professionals. Employees in fields that involve physical activity are relatively safe, probably because robots are not currently advanced enough to replace most jobs that involve physical interaction.

As AI becomes more capable, should the goal really be to gradually replace with machines everyone doing intellectual work, including programmers, writers, engineers, scientists, architects, doctors, lawyers, accountants, educators, etc.? There seems to be several problems with this "plan." Why would we want to have our best and brightest unemployed and not contributing to society? Companies might get an initial boost in earnings because their labor costs are reduced, but eventually few people will have any money to purchase the products produced by the companies that had laid them off. And do we really want to become the civilization described in E. M. Forster's "The Machine Stops," in which nobody knew how to fix "The Machine" as it gradually decayed and quit [66]?

In the above list, skilled laborers, such as pilots, science and medical technicians, electricians, mechanics, machinists, HVAC technicians, and carpenters should be included because these jobs often require detailed planning, precision work, math, and troubleshooting. However, numerous low-skilled jobs exist which involve moving boxes and equipment, scanning bar codes, driving vehicles, and cooking and serving fast food. In many states, the most common jobs are fast food workers, cashiers, and freight movers [67]. When we consider the amazing inventions that have been created and accomplishments that have been achieved by people, it demonstrates the incredible capacity of the human brain when it is trained and utilized to its highest potential. It seems like an extreme waste when a brain that could have had the potential to find the next cure for cancer is instead spending every day deciding how to move boxes from one place to another.

Another significant concern is the hazardous jobs that some people have, such as land and sea rescuers, underwater welders, astronauts, operators producing hazardous chemicals, underground miners, storm repair electricians for power companies, firefighters, policeman, soldiers, storm chasers, etc. Many of these employees probably enjoy this type of work because it provides a feeling of importance. However, I believe that there are many cases in which the risk is too high and that it would be better if we could avoid placing humans in harm's way.

These points lead me to believe that the implementation of AI should be utilized in the following ways:

- As assistants, and not replacements, for knowledge workers. Many people are already using AI in this way to enhance their capabilities, and this should continue. This allows people to accomplish more, but also to solve problems of higher complexity than they might have been able to do without the use of AI.
- To do mundane work, as described above. This might involve the replacement of a human that is moving boxes with a humanoid robot, but it could also apply to an electrician that has a humanoid robot as an assistant to do the mundane tasks that he/she usually has to do, such as fetching wire and tools, moving other equipment, preparing a location, cleaning up, etc.
- To do hazardous work. As for mundane work, this could involve the complete replacement of a human with a robot but might also involve the use of a robot assistant that does the most dangerous aspects of the work.

Implementing these ideas would require changes to the education system. Currently, most teachers do their best to educate all of their students to the highest level that they can. Some students do very well, while others do not grasp the material to the same degree. The result is a wide range of intellectual abilities. However, this is currently acceptable because there are a wide range of jobs requiring a wide range of mental abilities, from biochemical researcher to package delivery.

But with the suggested employment changes discussed above regarding AI implementation, humans would no longer be needed for lower skilled jobs. And everyone else would be operating at a higher intellectual level due to the AI assistants. Therefore, the education system would need modifications to ensure that all students succeeded to at least a minimum level that is higher than the current minimum level. Students that only achieve at the bare minimum level academically, may currently be able to find a menial labor job, but will not be able to once AI Entities have been deployed to those positions.

I want to close on a philosophical note. If we are able to create AI with AGI or ASI that is conscious and self-aware, and even has some form of emotions, we will have created Entities that are very human-like, essentially humans with silicon minds. Should they be given rights? Maybe equal to the rights that humans have? Will we need to be careful that they are not treated like the next incarnation of slaves? These types of questions have already been asked numerous times, but they have always been hypothetical because we have not had any AI Entities operating at the level discussed here. Of course, maybe we will decide that we do not want AI Entities that are conscious and we can avoid this messy issue. However, if consciousness and self-awareness can be implemented in an AI, someone will do it. So, we need to anticipate this development and be ready to decide how to proceed.

Bibliography

[1] F. Company, "Andrew Ng Says AGI is Decades Away," Inc., [Online]. Available: https://www.inc.com/fast-company-2/andrew-ng-agi-artificial-general-intelligence-ai-bubble-risk-training-layer/91310210?utm_source=flipboard&utm_content=topic/business. [Accessed 9 March 2026].

[2] G. Marcus, "BREAKING: LLM "reasoning" continues to be deeply flawed," Marcus on AI, [Online]. Available: https://garymarcus.substack.com/p/breaking-llm-reasoning-continues. [Accessed 9 March 2026].

[3] C. Metz, "An A.I. Pioneer Warns the Tech 'Herd' is Marching into a Dead End," The New York Times, [Online]. Available: https://www.nytimes.com/2026/01/26/technology/an-ai-pioneer-warns-the-tech-herd-is-marching-into-a-dead-end.html. [Accessed 9 March 2026].

[4] M. Heikkila and W. D. Heaven, "Yann LeCun has a bold new vision for the future of AI," MIT Technology Review, 24 June 2022. [Online]. Available: https://www.technologyreview.com/2022/06/24/1054817/yann-lecun-bold-new-vision-future-ai-deep-learning-meta/?gad_source=1&gad_campaignid=20737314952&gclid=CjwKCAiAs4HMBhBJEiwACrfNZU00u5-V_jZJElZuVG8ogDBdY1QJsuzULCh7kuima5DWbpC8HcKOwRoCVQQQAvD_BwE.

[5] "Artificial General Intelligence," Wikipedia, [Online]. Available: https://en.wikipedia.org/wiki/Artificial_general_intelligence. [Accessed 13 March 2026].

[6] S. Uboldi, A. Bortolotti, G. Candeloro, A. Marasco, F. Sardella, M. Tartari and P. L. Sacco, "From Touch to Mental Imagery: The Embodied Aesthetic Experience of Late-Blind People Engaged in the Tactile Exploration of Enrico Castellani's Pseudo-Braille Surface," *Culture, Medicine, and Psychiatry,* vol. 49, no. 3, pp. 725-764, 10 March 2025.

[7] "Hierarchy," Wikipedia, [Online]. Available: https://en.wikipedia.org/wiki/Hierarchy.

[8] P. Reber, "Ask the Brains," *Scientific American MIND,* 2010.

[9] J. S. Mill, "A System of Logic: Ratiocinative and Inductive, Presenting a Connected View of the Principles of Evidence and the Methods of Scientific Investigation," First ed., vol. II, 1843, p. 245.

[10] "Causal Reasoning," Wikipedia, [Online]. Available: https://en.wikipedia.org/wiki/Causal_reasoning.

[11] "Brain Metrics Initiative," [Online]. Available: https://www.test-iq.org/the-bmi-certified-iq-test/.

[12] O. C.-F. M. P. d. A. R. R. d. P. S. a. R. d. C. G. Karina Matugi, "Vapor-Liquid Equilibrium Calculation for Simulation of Bioethanol Concentration from Sugarcane," *Brazilian Journal of Chemical Engineering,* vol. 35, no. 2, pp. 341-352, 2018.

[13] D. F. Riddle, Calculus and Analytic Geometry, Belmont, CA: Wadsworth Publishing Company, Inc., 1974, p. 341.

[14] "Maslow's Hierarchy of Needs," Wikipedia, [Online]. Available: https://en.wikipedia.org/wiki/Maslow%27s_hierarchy_of_needs. [Accessed 24 January 2024].

[15] S. Mcleod, "Maslow's Heirarchy of Needs," SimplyPsychology, 24 January 2024. [Online]. Available: https://www.simplypsychology.org/maslow.html.

[16] A. H. Maslow, "A Theory of Human Motivation," *Psychological Review,* vol. 50, pp. 370-396, 1943.

[17] "Need," Wikipedia, [Online]. Available: https://en.wikipedia.org/wiki/Need. [Accessed 24 January 2024].

[18] "Emotion," Wikipedia, [Online]. Available: https://en.wikipedia.org/wiki/Emotion. [Accessed 4 February 2024].

[19] "Emotion Classification," Wikipedia, [Online]. Available: https://en.wikipedia.org/wiki/Emotion_classification. [Accessed 10 February 2024].

[20] I. Asimov, "Runaround," in *I, Robot*, Gnome Press, 1950.

[21] "Uncertainty," Merriam-Webster, [Online]. Available: https://www.merriam-webster.com/dictionary/uncertainty. [Accessed 25 April 2026].

[22] "Probability," Merriam-Webster, [Online]. Available: https://www.merriam-webster.com/dictionary/probability. [Accessed 25 April 2026].

[23] "15 Phrases for Certainty and Probability," Oxford Language Club, [Online]. Available: https://oxfordlanguageclub.com/page/blog/15-phrases-for-certainty-probability. [Accessed 23 March 2024].

[24] K. Boogard, "What is a Risk Matrix?," Wrike, [Online]. Available: https://www.wrike.com/blog/what-is-risk-matrix/. [Accessed 18 February 2024].

[25] A. Stein, "English Words with Multiple Meanings," Lingoda, 1 June 2021. [Online]. Available: https://blog.lingoda.com/en/english-words-that-mean-something-different/.

[26] J. Rubin, "How punctuation (somehow) plays a role in Trump's Jan. 6 case," MSNBC, 28 Nov. 2023. [Online]. Available: https://www.msnbc.com/deadline-white-house/deadline-legal-blog/punctuation-somehow-plays-role-trumps-jan-6-case-rcna126989.

[27] A. O'Reilly, "What is an Idiom? Definition and Examples," Grammarly, 1 November 2022. [Online]. Available: https://www.grammarly.com/blog/idiom/.

[28] R. Nordquist, "The 9 Parts of Speech: Definitions and Examples," ThoughtCo., 11 March 2020. [Online]. Available: https://www.thoughtco.com/part-of-speech-english-grammar-1691590.

[29] "Merriam-Webster Help," Merriam-Webster, [Online]. Available: https://www.merriam-webster.com/help/faq-how-many-english-words. [Accessed 19 March 2024].

[30] "Prepositions of Place," LanGeek, [Online]. Available: https://langeek.co/en/grammar/course/69/prepositions-of-place. [Accessed 23 March 2024].

[31] "Parts of Speech: Pronouns," Scribendi, [Online]. Available: https://www.scribendi.com/academy/articles/parts_of_speech_pronouns.en.html. [Accessed 10 November 2024].

[32] "Parts of Speech: Adverbs," Scribendi, [Online]. Available: https://www.scribendi.com/academy/articles/all_about_adverbs.en.html. [Accessed 16 November 2024].

[33] "Prepositions: Definition and Examples," Grammarly, [Online]. Available: https://www.grammarly.com/blog/parts-of-speech/prepositions/. [Accessed 17 November 2024].

[34] Q. B. Institute, "The Brain - Memory and Learning," [Online]. Available: https://qbi.uq.edu.au/brain-basics/memory. [Accessed January 2024].

[35] "Memory," Wikipedia, 23 December 2023. [Online]. Available: https://en.wikipedia.org/wiki/Memory.

[36] "Organization of Long-Term Memory," The Peak Performance Center, 2024. [Online]. Available: https://thepeakperformancecenter.com/educational-learning/learning/memory/stages-of-memory/organization-long-term-memory/.

[37] S. Pappas, "How does the brain store memories," LiveScience, 27 September 2022. [Online]. Available: https://www.livescience.com/how-the-brain-stores-memories.

[38] A. F. de Sousa, A. Chowdhury and A. J. Siva, "Dimensions and mechanisms of memory organization," *Neuon,* vol. 109, no. 17, pp. 2649-2662, 2021.

[39] "Semantic Memory," Wikipedia, [Online]. Available: https://en.wikipedia.org/wiki/Semantic_memory. [Accessed 2 January 2024].

[40] R. J. Branaghan and S. Lafko, "Chapter 20 - Cognitive Ergonomics," in *Clinical Engineering Handbook*, Second ed., E. Iadanza, Ed., Academic Press, 2019, pp. 847-851.

[41] M. Naim, M. Katkov, S. Recanatesi and M. Tsodyks, "Emergence of hierarchical organization in memory for random materials," *Scienific Reports,* vol. 9, no. 10448, 2019.

[42] S. Pinker, Words and Rules, New York: Perennial, 1999.

[43] "Percy Spencer," Wikipedia, [Online]. Available: https://en.wikipedia.org/wiki/Percy_Spencer. [Accessed 7 December 2025].

[44] "Wright Brothers," Wikipedia, [Online]. Available: https://en.wikipedia.org/wiki/Wright_brothers. [Accessed 7 December 2025].

[45] "Proportional–integral–derivative controller," Wikipedia, [Online]. Available: https://en.wikipedia.org/wiki/Proportional%E2%80%93integral%E2%80%93derivative_controller. [Accessed 7 December 2025].

[46] G. Marcus and E. Davis, Rebooting AI: Building Artificial Intelligence We Can Trust, First ed., New York: Pantheon Books, 2019, p. 113.

[47] Merriam-Webster, "Consciousness," Wikipedia, [Online]. Available: https://www.merriam-webster.com/dictionary/consciousness. [Accessed 5 February 2026].

[48] V. Caston, "Aristotle on Consciousness," *Mind,* vol. 111, no. 444, pp. 751-814, 2002.

[49] "John Locke," Wikipedia, [Online]. Available: https://en.wikipedia.org/wiki/John_Locke. [Accessed 5 February 2026].

[50] J. Locke, An Essay Concerning Hunan Understanding, vol. 4, London, 1689.

[51] "The Consciousness Secret: How John Locke Defined the Self," Philosopheasy, 1 May 2025. [Online]. Available: https://www.philosopheasy.com/p/the-consciousness-secret-how-john.

[52] "David Hume," Wikipedia, [Online]. Available: https://en.wikipedia.org/wiki/David_Hume. [Accessed 7 February 2026].

[53] "A Treatise of Human Nature," Wikipedia, [Online]. Available: https://en.wikipedia.org/wiki/A_Treatise_of_Human_Nature. [Accessed 7 February 2026].

[54] "David Hume - Is there a Self?," Valkyrist, [Online]. Available: https://valkyrist.wordpress.com/2014/01/15/david-hume-is-there-a-self/. [Accessed 7 February 2026].

[55] "Edmund Husserl," Wikipedia, [Online]. Available: https://en.wikipedia.org/wiki/Edmund_Husserl. [Accessed 7 February 2026].

[56] "Phenomenology," Wikipedia, [Online]. Available: https://en.wikipedia.org/wiki/Phenomenology_(philosophy). [Accessed 7 February 2026].

[57] E. Husserl, Logical Investigations, Germany: M. Niemeyer, Routledge & Kegan Paul Ltd, 1901.

[58] "Existentialism," Wikipedia, [Online]. Available: https://en.wikipedia.org/wiki/Existentialism. [Accessed 7 February 2026].

[59] "Jean-Paul Sartre," Wikipedia, [Online]. Available: https://en.wikipedia.org/wiki/Jean-Paul_Sartre. [Accessed 7 February 2026].

[60] J.-P. Sartre, Being and Nothingness, France: Éditions Gallimard, Philosophical Library, 1943.

[61] R. Lewis M.D., "An Overview of the Leading Theories of Consciousness," Psychology Today, 25 November 2023. [Online]. Available: https://www.psychologytoday.com/us/blog/finding-purpose/202308/an-overview-of-the-leading-theories-of-consciousness.

[62] "What is the Memory Capacity of a Human Brain," Clinical Neurology Specialists, [Online]. Available: https://www.cnsnevada.com/what-is-the-memory-capacity-of-a-human-brain/. [Accessed 13 February 2026].

[63] E. Tufte, "Retina communicates to brain at 10 million bits per second: Implications for evidence displays?," Graphics Press, 26 July 2006. [Online].

[64] B. A. Sauerbrei and J. A. Pruszynski, "The brain works at more than 10 bits per second," *Nature Neuroscience,* vol. 28, no. 7, pp. 1365-1366, 2025.

[65] I. Asimov, I, Robot, United States: Gnome Press, 1950.

[66] E. M. Forster, The Machine Stops, Oxford and Cambridge Review, 1909.

[67] D. Neufeld and A. Duo, "Mapped: The Most Common Job in Every U.S. Stage," Visual Capitalist, 26 May 2026. [Online]. Available: https://www.visualcapitalist.com/most-common-job-in-every-u-s-state/.

[68] "Common Prepositions," Excelsior Online Writing Lab, [Online]. Available: https://owl.excelsior.edu/grammar-essentials/parts-of-speech/prepositions/common-prepositions/. [Accessed 17 November 2024].

[69] "What is a Preposition? How to Use It & Examples," Scribendi, [Online]. Available: https://www.scribendi.com/academy/articles/what_is_a_preposition.en.html. [Accessed 17 November 2024].

www.ingramcontent.com/pod-product-compliance
Lightning Source LLC
LaVergne TN
LVHW090348240926
848801LV00017B/184/J